*This book is a must read for those who want to know the how and why of reversing the destruction of our world. As a pastor and a Creation Care leader, I was fascinated and educated. The principles and practical action items are simple and inspiring. Simply read, pray, and follow directions!*

Dr. Joel C. Hunter
Senior Pastor, Northland - A Church Distributed

*Using Psalm 8 as a simple template Edward Brown presents the glory and fantastic scale of God's creation. He then emphasizes that the special status humans have been given under God to have dominion over the earth comes with an awesome responsibility. But humans are raping the earth instead of caring for it. Brown eloquently describes what God expects from us and goes on to present the enormous challenge to all peoples but especially the world's Christians squarely to face the demands of caring for creation and all that that means. He urges us to take a global perspective and get real and active realizing that we don't have to do it on our own. God is offering to partner with us in furthering His kingdom and bringing about His will on Earth as it is in heaven.*

Sir John Houghton
Director of the John Ray Initiative (UK)
Former co-chair of the IPCC Scientific Working Group

*Ed Brown has done it again. He has written an important book for Christians who are serious about their faith. He has written another book that I want to read again. He challenges Christians to be serious about stewarding God's creation. The book reveals the author's deep reverence for the Word of God, the work of God, and the potential of the people of God. He challenges Christians to start with right thinking about Creation and then move to right living. He does this by taking us deep into the nature of God and God's intentions for all that is made. Ed makes clear that proper stewardship of Creation is not an option that Christians can decide to add to their value set or not. Rather, caring for all that God has made is essential to both participating in and building the Kingdom of God. What emerges is a marvelous call for Christians to live up to who God has made us to be in Christ – partners in the ministry of reconciling all things to Christ, and drawing clear and unambiguous attention to the importance of properly stewarding Creation. This, then, is a book for Christians of all ages, vocations, and backgrounds serious about their faith.*

Director, Natural E₁

*We're running this world into the ground and we desperately need another way forward. As Ed Brown rightly puts it, we need God's way. When Heaven and Nature Sing is for all those who are ready to surrender our goals for God's goals, and exchange our plan for God's plan. I'm grateful for this empowering and hope-filled book, just as I'm grateful for Ed Brown's visionary leadership and faithful ministry.*

Ben Lowe
Director of Young Evangelicals for Climate Action
Author of *Green Revolution*

*We live in a world of beauty, yet brokenness. This world is not how God intended it to be: people have preferred to put themselves in charge rather than to live under the rule of their creator God. They have ravaged the environment for their own gain. This little book shows how living out of kilter with God has led to the pervasive environmental and personal problems we all face. Ed Brown develops six helpful biblical principles for how we might behave so as to reflect God's goals for the world. If Christians consistently and practically applied these principles, it would revolutionize our world. Recommended for everyone who takes seriously their responsibility for their lifestyles, daily choices and behaviour as part of their worship of God.*

Bob White
Professor of Geophysics, Cambridge University
Director of the Faraday Institute for Science & Religion
Fellow of The Royal Society

# WHEN HEAVEN AND NATURE SING

EXPLORING GOD'S GOALS FOR HIS PEOPLE AND HIS WORLD

EDWARD R. BROWN

*For Paul & Lynn —*
*with much appreciation*
*for your friendship &*
*generosity*

*Ed*
*4/4/13*

Doorlight Publications,
South Hadley, Massachusetts

*Doorlight Publications*
*www.doorlightpubs.com*

*ISBN 0-9838653-1-0*

*ISBN13 978-0-9838653-1-5*

*Joy to the world, the Lord is come!*
*Let earth receive her King;*
*Let every heart prepare Him room,*
*And Heaven and nature sing,*
*And Heaven and nature sing,*
*And Heaven, and Heaven, and nature sing.*

*Joy to the world, the Savior reigns!*
*Let men their songs employ;*
*While fields and floods, rocks, hills and plains*
*Repeat the sounding joy,*
*Repeat the sounding joy,*
*Repeat, repeat, the sounding joy.*

*No more let sins and sorrows grow,*
*Nor thorns infest the ground;*
*He comes to make His blessings flow*
*Far as the curse is found,*
*Far as the curse is found,*
*Far as, far as, the curse is found.*

*He rules the world with truth and grace,*
*And makes the nations prove*
*The glories of His righteousness,*
*And wonders of His love,*
*And wonders of His love,*
*And wonders, wonders, of His love.*

Isaac Watts, 1719

*To Vern,*
*Lowell,*
*Tom*
*and Ben*
*fellow travelers in a great adventure!*

# CONTENTS

# FOREWORD

Devastating storms, droughts, and fires ravage larger and larger swaths of Earth. The media call these "extreme weather events." Politicians and others debate the causes. Some environmentalists say it is now too late to save our planet from disaster.

Should Christians give up hope? Many people who are deeply concerned about the environment are starting to. Responsible climatologists' projections are not encouraging. Some business and government leaders say: We simply need to adapt to a hotter climate; turn potential disaster into economic opportunity. Turn the dust of drought to gold.

Meanwhile, many Christians are confused. I hear three main reactions. One: God is in control, so we don't have to worry; the whole world is in his hands. Two: There really is no problem; alarms about the environment are politically motivated and just plain wrong. Or three: Yes, we face possible environmental disaster, but this is such a stupendous, complex matter that we can't really deal with it. So we do nothing.

Ed Brown's fine new book When Heaven and Nature Sing is a splash of refreshing water in the face of such responses. Brown has looked carefully at the data, but he's also looked faithfully at the Bible. He's banking on God's goodness and his promises and power, and so should we. The six key principles he outlines offer timely help to the church and its mission today.

The earth has now entered the Anthropocene Age, as Ed notes. We must rethink the idea that the earth is so vast that humanity cannot change it very much. This is simply not true. We are damaging the world, fouling the nest, poisoning the garden (often literally) upon which our human life under God depends.

This book is packed full of wisdom—wisdom that is both biblically based and ecologically sound. We can and should pay careful attention to the "cosmic golden rule" of which the author speaks. God is sovereign; we are responsible.

I especially appreciate the author's emphasis on worship, on the kingdom of God, and on what he calls "biospherical proclamation – the proclamation of God's name through the care and nurturing of his creation, in particular the biosphere, the realm of living organisms." This is both biblical and strategically important. We need to understand and put into practice this key insight: "Biospherical proclamation requires the preservation and flourishing of creation as a central part of the proclamation task."

"Things go better when we do them God's way." Can there be any doubt about that? As Christian disciples concerned about God, about the church and witness, and about our hurting world, we can act on that assurance. Empowered by God's Spirit, we can carry the full dynamic of the Good News of Jesus Christ into the very thick of today's environmental crisis and be the gracious healers and change agents that God intends.

We have hope, because Jesus lives and his Spirit works in the world. Time is short, but God is strong. I am praying this book will be used by God's Spirit both to inflame hope and to prompt and guide intelligent, faithful action, to the glory of God and the flourishing of the earth—both its human and nonhuman inhabitants.

*Howard A. Snyder*
*Author with Joel Scandrett of*
Salvation Means Creation Healed

# INTRODUCTION

Five years ago I discovered E. O. Wilson's little book, *The Creation*, in a local bookstore. Wilson, Harvard Biologist and a prominent leader of the modern environmental movement, addressed a plea to a hypothetical pastor: "Dear Pastor, we scientists need your help." Wilson was right. The world desperately needs the church to join in a response to the environmental crisis. That was the nudge that I needed to begin writing *Our Father's World: Mobilizing the Church to Care for Creation*. And, as I showed in that book, the church is not only needed, she is uniquely suited to lead that response.

Five years later, where are we?

*The environmental crisis continues unabated.* In spite of occasional breakthroughs here and there the general trend is an acceleration of degradation. Our ecological house is falling apart. There is no need to elaborate. Pick up any newspaper or scan the environment section of any major news website and you will read the same story. Or talk to someone from Kenya or the Philippines or Haiti. In these and any number of other countries human suffering and environmental degradation are increasingly linked.

I began *Our Father's World* with a description of the scene outside my dining room window, where an apparently pristine restored prairie conceals an old city landfill. I suggested that this was a metaphor for the environmental crisis today: covered up, often invisible but still dangerous. If we were to update that analogy now, I would say that a third of the prairie has been peeled back. The foul garbage beneath is breaking through in many places. The environmental crisis today is a lot more obvious than it was just five years ago.

Have we made any progress in figuring out what to do about it?

*Responses are fragmented.* A difficult political climate combined with a severe economic recession has pushed environmental concerns to the back burner both for the general public and within the church. When you don't have a job, or when you are looking at the loss of your home, it's hard to get excited about recycling or using alternative energy. This is understandable. But it can have tragic results. Like a parent forced to neglect her children's health by skimping on milk and vaccinations to pay the rent, it might seem un-avoidable – no, it might be unavoidable – but there is a heavy price to pay. As with children's teeth damaged by poor nutrition, environmental conse-quences won't wait while we put the economy back together. Those of us tempted to cheer for jobs over environmental action should be aware of the long-term effects of this kind of strategy.

*Environmentalists are discouraged.* Environmental and economic trends are accelerating, driven by an inexorable increase in the number of human beings on the planet. Prominent writers and thinkers are starting to look beyond the present crisis toward what they see as an inevitable crash and are asking how people and communities can survive what appears to be in-evitable. This sounds like the stuff of doomsday websites and people who study ancient South American calendars, but these individuals are smart and well respected. Folks like Bill McKibben, who is convinced that our planet has already been so changed that it needs a new name: He calls it Planet Eaarth and his book by that name is worth your time. James "Gus" Speth, who has spent much of his career in the White House, argues that our entire market capitalist system is doomed and needs to be replaced. The standard watchword of the environmental movement, 'sustainability', is being pushed aside by a new word, 'resilience'. It is too late to talk of sus-taining our present course of action. Instead we need to build a society that is resilient – capable of withstanding the inevitable challenges caused by centuries of environmental abuse.

Reading these authors, I feel like a passenger in a damaged airplane at the moment when the pilots finally conclude that a crash landing is inevi-table. Focus suddenly shifts from keeping the plane in the air to preparing for impact. That's the position McKibben, Speth and many others think we are in right now.

*But it is not yet time to give up hope.* Encouraging changes are in the wind. The environmental movement as a whole is more people-centered than it used to be. I live in Madison Wisconsin, which claims Gaylord Nelson, founder of the modern Earth Day movement, as well as John

Muir, Aldo Leopold and Sigurd Olsen, all giants in the environmental movement from earlier generations, and Cal DeWitt in our own day. This rich history means that major environmental conferences are regularly held here, and I have been able to attend several of these in recent years. I almost always come away impressed by what seems to be a new awareness that people are important. There is a new understanding that we cannot solve environmental problems without also solving the problems of the people who live there. The environmental crisis isn't just about spotted owls or polar bears any more, it's about us.

*I see a merging of environmental and social justice concerns.* Paul Hawken calls the massive efforts of millions of people around the world to work for improved environmental conditions and justice for oppressed peoples a "movement without a name." He considers it one of the greatest movements in human history (Paul Hawken, Blessed Unrest). I'm not sure he's right. The Church of Jesus Christ, properly understood, is surely the greatest movement in history. Hawken is wrong in thinking that we human beings can somehow pull ourselves up by our bootstraps. But he is right that a new social and environmental awareness is gaining momentum around the world.

That is where this book starts.

The central premise of *Our Father's World* was that our main problem is spiritual. A problem that began by disobeying the Creator is not going to be solved by reorganizing our economies, by inventing new technologies or by marching in the streets. We are doomed until we restore our fundamental relationship with God. Hoping for the best isn't going to go anywhere.

This book goes a step further. Having a restored relationship with God, can we have a world that reflects God's goals rather than our own? And what does that mean? Is it even possible to know what God's goals for the world (and for us) are?

If you have read *Our Father's World*, you could read this as a sequel. It is that. But you will find that we're working with some very different material this time. Creation care is an important theme in this study, but this is really not just a creation care book. The implications of this study are broad and deep and eventually touch on every possible aspect of human society and civilization.

One of my friends asked, "Are you writing a manifesto?" I didn't set out to do so. But if the ideas I've articulated are taken seriously, they will force us to rethink everything we do as human beings. Everything. Does this rise to the level of a manifesto? If so, is it a manifesto worth adopting? Our problems (plural) are not just polluted rivers or a changing climate or an unjust economic system or ineffective schools or massive public debt or dysfunctional political institutions. Our (singular, upper-case "P") Problem is that more than ten thousand years after the Garden of Eden and more than 2000 years since Jesus came, lived, died and rose from the dead we are still living in our sins – collectively and individually. And we are learning that sin has consequences.

The good news is that we don't need to keep living in our sins, individually or collectively. The redemption that came with Jesus applies not only to our individual sins and sorrows, but to those of our communities and our society as well. No, we can't build a perfect society now, any more than you can live a perfect day or week in your own life. But we can look toward the day when Jesus really will rule the world with truth and grace, and heaven and nature will sing and sing and sing. And we can understand God's goals and work toward them as we shape this world that we live in.

I want to thank some of those who have made this book possible. My wife Susanna was patient and encouraging throughout. My now grown children were invaluable in their initial reactions to the first draft. Vern and Lowell made very helpful suggestions as the manuscript progressed. Tom and Dan, the driving forces behind Doorlight Publications, showed once again that it is possible to bring a book to print in less time than I would have thought possible. And finally, I would like to thank those who have stood with me and Susanna as financial partners in our ministry, some for many, many years. You know who you are – and you should know that it is because of you that we are able to do what we do.

Thank you!

*Ed Brown*
*Madison, Wisconsin*
*September 2012*

# Part I
# Who is Running this Place?

IN WHICH WE EXPLORE QUESTIONS OF AUTHORITY,
DOMINION AND GOALS.

# 1

## WHO'S RUNNING THIS PLACE, ANYWAY?

My youngest daughter would like to run the world. It started when, as a student at the University of Minnesota, she spent a summer working at the Minnesota Zoo. She recorded the experience on her first blog with a title that played on one of her favorite Dr. Seuss books: "If Amy Ran The Zoo."

A few months later she went on a semester abroad, and the blog became "If Amy Ran Senegal." Returning from that experience, with graduation coming and the whole world in front of her, the blog morphed into the present "If Amy Ran the World..."

Amy is not alone. Whether it's the company where we work, the community we live in, or national politics, we all have times when we want to say, "Who's running this place, anyway? If only I were in charge..."

So who is running the world?

### Somebody's behind that curtain!

When we are children, the world just is. Everything around us feels permanent. My house, my school, my street have always been there and always will be. Whether it is mountains or a sliding hill in the park, Niagara Falls or the fountain in front of City Hall, the Grand Canyon or the gulley at the end of the block, as children we don't think about how these things came to be. They just are.

As we grow older, we learn that the world we live in is constantly changing. It is the very nature of things. A pet dies. A familiar tree blows over. An empty lot becomes a new apartment complex. The world is not the static place we thought it was.

Still later we learn that the world has a history. Things we see now haven't always been there. My house used to be a forest. My children's school was a pasture. The sliding hill in the park covers an old city dump; that gulley is a result of poor storm water management; the fountain is there because my grandparents and others raised money for it.

We also learn that the world didn't just 'happen'. It is in a constant process of transformation, by nature and by people. Like Dorothy in the Wizard of Oz we learn that there is someone behind the curtain. The world we live in is a result of human manipulation and influence. We discover that, in reality, people are running the world.

It is true that ultimately God is the one running the world, and I don't mean to ignore that important fact. But it is also true, as we are going to see, that in the day-to-day events that make up our world, God is not the most immediate and visible cause. People are. God's ultimate control does not in any way cancel the reality that we human beings are shaping and molding the world in which we live.

As I write this, I am looking out the window at a street lined with trees turning gold, as they do in late September. This neighborhood feels permanent, but the houses are only about twenty years old. I am almost three times as old as the street.

This street is where it is because a developer bought a farm and decided that there would be a street here. The fact that I can see eight golden trees, and not three or twenty is because someone decided how many trees would be planted, and what kind – and someone else went out and put the seedlings in the ground. The "nature" that I see out my window was made by God, but was put there by people!

Out the other side of the house, I am looking at the restored-prairie-that-is-really-a-former-city-dump, that I described in *Our Father's World*. What appears to be innocent and even beautiful greenspace, nicely placed between neighborhoods, hides environmental degradation.

Everything I see has been shaped, reshaped and molded by human beings. I could take you on a five-minute drive and show you miles of development that has happened in just the last 10 years, within my own short time in this city.

But Madison is not exactly wilderness. What if we go to the real wilderness, the rain forest, or the high Himalayas. Surely there you will be out of reach of humanity! Surely there you can see 'unspoiled' wilderness! We small human beings don't really run these parts of the world, do we?

## Welcome to the Anthropocene

Bill McKibben, one of today's best-known environmentalists, created quite a stir when his book *The End of Nature* came out several years ago. The book was an early call to pay attention to the problems the human race would be facing as a result of anthropogenic (human-induced) climate change. But his central thesis was bigger. McKibben argued that "undisturbed nature" is a myth. It is not just my neighborhood that has been shaped by human activity, McKibben argues. Every part, every square inch of our planet has been impacted and changed by what we do. There is no such thing as "nature" any more:

> What mattered to me most was the inference I drew from [climate] science: that for the first time human beings had become so large that they altered everything around us. That we had ended nature as an independent force, that our appetites and desires could now be read in every cubic meter of air, in every increment on the thermometer...
>
> ...we are no longer able to think of ourselves as a species tossed about by larger forces – now we are those larger forces. Hurricanes and thunderstorms and tornadoes become not acts of God but acts of man. That was what I meant by the end of nature. (McKibben 2006: xviii)

Geologists use terms like Paleozoic, Mesozoic and Cenozoic to describe ancient geological eras lasting millions of years; Pliocene, Pleistocene and Holocene mark shorter (but still very long) geological epochs. Up to now there has been general agreement that we are living in the Holocene, Greek for "Most Recent Era". But scientists are in the process of adding a new epoch to this list: "Welcome to the Anthropocene", says a recent cover of the Economist magazine. We are now in the age of man.

Almost 90% of the world's plant activity, by some estimates, is to be found in ecosystems where humans play a significant role. Although farms have changed the world for millennia, the Anthropocene advent of fossil fuels, scientific breeding and, most of all, artificial nitrogen fertiliser has vastly increased agriculture's power. The relevance of wilderness to our world has shrunk in the face of this onslaught. The sheer amount of biomass now walking around the planet in the form of humans and livestock handily outweighs that of all other large animals. The world's ecosystems are dominated by an increasingly homogenous and limited suite of cosmopolitan crops, livestock and creatures that get on well in environments dominated by humans. Creatures less useful or adaptable get short shrift: the extinction rate is running far higher than during normal geological periods. (*The Economist* May 26, 2011 )

Question: Who is running this place?

Answer: We are.

Next Question: How are we doing?

**A performance review?**

In May 2009 Paul Hawken spoke to the graduating class of the University of Portland. His short talk went viral:

Class of 2009: You are going to have to figure out what it means to be a human being on earth at a time when every living system is declining, and the rate of decline is accelerating. Kind of a mind-boggling situation... but not one peer-reviewed paper published in the last thirty years can refute that statement. Basically, civilization needs a new operating system, you are the programmers, and we need it within a few decades ...We have tens of thousands of abandoned homes without people and tens of thousands of abandoned people without homes. We have failed bankers advising failed regulators on how to save failed assets. We are the only species on the planet without full employment. Brilliant. We have an economy that tells us that it is cheaper to destroy earth in real time rather than renew, restore, and sustain it. You can print money to bail out a bank but you can't print life to bail out a planet. At present we are stealing the future, selling it in the present, and calling it gross domestic product. (Hawken: 2009)

While I was writing this book Japan suffered and was still recovering from the triple tragedy of earthquake, tsunami and nuclear meltdown. The Gulf coast of the United States was still scarred from an oil spill a year earlier. A housing crisis that began three years before was nowhere near resolution. A famine in the horn of Africa, unnoticed by western media, gradually spread to include most of North Africa. The economies of Europe were in a shambles. Climate-change related weather difficulties stalked the globe, from drought and wildfires in Texas to a second straight year of massive flooding in Pakistan to the summer of 2012 when my own hometown in Wisconsin broke heat records set in the 19th century on multiple days, often by 5 degrees or more. While there was more wealth circulating the globe than ever before, there were also more slaves than there ever have been, and even those who had money were measurably less happy than the "poorer" generations that preceded them. Possibly not unrelated, late in 2011 the world symbolically welcomed the 7 billionth baby into the world. The human population has almost tripled in my lifetime.

I could give you pages of data and stories about all the ways that we're messing up God's world, but I don't think you need a lot of proof: You live in the same world I do, you see the same things going on in your community, your state, your country that I see. And there are many other books and countless websites that can give you the data if you need it.

No. If I were to give you 250 pages, it would all come down to the same conclusion: We human beings are running God's world, and we're making a mess of it.

That sets up our central proposition: It is possible to do things differently.

# 2

## Doing Things God's Way

I'm an optimist. Sometimes my wife thinks I'm a bit too much of an optimist, but that's a story for another day or another book. However, I really do think we can run God's world differently than we have been doing. We are not doomed to repeat the errors of the past. We don't have to have nuclear plants on earthquake faults. We don't have to destroy the ecosystems that feed our bodies in order to satisfy our appetite for energy. It is not necessary to have an economic system so unbalanced that some farmers destroy food they can't sell while millions starve elsewhere. And it should not be inevitable that the needs of our lives today will destroy the climate system on which the lives of future generations will depend.

We can do things differently.

### No help from history

But hang on, Ed. Lots of people have tried to "do things differently" in the past, and it usually didn't work out.

You're right. That little phrase, "Let's do things differently" captures much of human history.

"Let's do things differently," our ancestors said, and went off and invented new political systems, new economic systems, new social structures, even new religions. And thus came communism, capitalism, socialism, and a host of other -isms.

Sometimes things got better. Sometimes worse. Most often things at street level didn't look very different than they looked before.

I saw this first-hand in Manila in February, 2011. For decades after World War II the Philippines, a former colony of the United States, suffered under the oppressive dictatorship of Ferdinand Marcos. In February 1986 Marcos was overthrown by a remarkable bloodless revolution that became known as the People Power Revolution. It became a model for a number of others that followed, Eastern Europe in 1989 and, some think, even the Arab Spring of 2011.

I happened to be in Manila on the very weekend that the entire country celebrated the twenty-fifth anniversary of People Power. One evening I attended the meeting of a monthly gathering of Filipino writers. This was a salon in the classical sense of the word, a gathering of intellectuals for mutually stimulating conversation.

The reason for the meeting was a book launch, but the main topic of conversation was the events of twenty-five years before. Many of those present, including the author of the book, had been involved in the uprising in various ways, and their memories were still vivid.

The conversation was lively, half in English and half in Tagalog: Filipinos switch so quickly between the two that you are often not sure which language you are listening to. That hardly mattered because an argument proposed in Tagalog would be countered in English, so even a non-Filipino was able to follow most of the debate. It was clear that the memories of a glorious past were tempered by a sense of disappointment and disillusionment in the present. The People Power revolution had been a success – Marcos was gone. But the real dreams of these reformers had not been fulfilled. They had not wanted merely to replace a dictator. They had wanted to remake their country. They had wanted to end corruption, poverty and disease, environmental degradation. None of this had happened. Corrupt dictators had been replaced by corruptly elected politicians. In the slums of Manila and in the countryside, nothing had changed.

In our own day we are watching countries in the Middle East go through a similar process in what has been called the "Arab Spring." Egypt is the most familiar case. It always seems to be the same story: The more things change, the more they stay the same.

**Maybe if we tried to do things God's way?**

We can do things differently, I said above. "Really?" you replied. There is nothing new under the sun. What is there that hasn't been tried – and failed – before?

> *All streams run to the sea,*
> *but the sea is not full;*
> *to the place where the streams flow,*
> *there they flow again.*
> *All things are full of weariness;*
> *a man cannot utter it;*
> *the eye is not satisfied with seeing,*
> *nor the ear filled with hearing.*
> *What has been is what will be,*
> *and what has been done is what will be done,*
> *and there is nothing new under the sun.*
> Ecclesiastes 1:7-9

But maybe there is something no one has ever really tried before. Maybe we haven't been trying to do things differently enough.

Let me take you on a brief detour to Africa to explain.

**Doing things God's way on the farm**

At Care of Creation our staff have been researching, testing and implementing a program called "Farming God's Way" on the slopes of the great Rift Valley in central Kenya.

The program is based on a simple premise: Things might go better on our farms if we try to do things the way God might do them, or "God's way". This may sound presumptuous – how can someone claim to know what God's way is? But listen to how the program got started.

In 1982 Brian Oldrieve became the manager of Hinton Estates, a large commercial farm in Zimbabwe. For his first two years he struggled with the same challenges as all of the other farms around: High input costs, damaging soil erosion, and rapid loss of soil fertility with serious declines in yields. Brian started to pray, asking God to show him how he could manage these fields, God's fields, in such circumstances.

Here's how one of Brian's colleagues describes the process:

Brian asked God to teach him how to overcome the difficulties he was encountering and so the discovery of Farming God's Way began. God began to speak to him about how he manages creation, where he never tills or inverts the soil and where he never destroys or buries decaying plant matter on the surface. God also spoke clearly of the stewardship and excellence farmers are called to adopt, by caring for and managing the land in a way that would ensure that farmlands could be passed on as an inheritance from one generation to another and bring glory to God. (*Farming God's Way Training Guide 11*)

Most of us aren't accustomed to hearing God "speak to us" quite this way, but don't let that distract you from the main point. Brian listened to God, and learned from God, whatever the actual form of communication was.

For Brian, doing things "God's Way" meant two things: to work in harmony with how God runs his world – the natural operations of God's creation as we can discover them; and to work according to the principles of the Word of God, the Bible, as we learn and understand them. Understand the world. Study the Word. Put them together and do the thing "God's Way".

### Understanding how God's world works

The first part of Brian's "discovery" was actually quite close to what research programs in the West had started calling conservation no-till agriculture. This method of farming has its roots in the work of Edward Faulkner in the mid-Twentieth Century. Faulkner's book, *Ploughman's Folly* (1943), generated a great deal of controversy. "No one," Faulkner wrote, "has ever advanced a scientific reason for plowing." (Faulkner 3). He advocated leaving the soil intact, rather than disturbing it with a plow. He placed a strong emphasis on soil health and on the use of a layer of organic compost. At the time, things were not going that well with traditional farming methods. In the U.S. the 1930's Dust Bowl was still a fresh memory. It was time to reconsider basic assumptions about how we do things. Others built on this early work, including essayist Wendell Berry:

An agriculture using nature, including human nature, as its measure would approach the world in the manner of a conversationalist. It would not impose its vision and its demands upon a world that it conceives of as a stockpile of raw material, inert and indifferent to any use that may be made of it... (Berry 1990: 208-9)

'Listen to the land!' seems to be one of the messages of Faulkner and Berry. Brian Oldrieve would agree, to a point. But there is more to Brian's approach, for Brian really wants us to listen to God.

### Listening to what the Word says

Remember, Brian was not just observing the world; he was also teaching people to obey God's Word, the Bible. Farming God's Way is an agricultural program that uses Christian discipleship, or a discipleship program that uses agriculture. Actually, it is both, and therein lies its genius.

Farming God's Way is presented in a kind of faith-language that resonates culturally with Africans: The fields are God's fields. We cover God's field with "God's blanket" (a layer of organic compost). We work hard, and with careful planning and discipline because this is how God wants his people to work. Brian learned to communicate a sophisticated scientific method of farming in culturally appropriate language that spoke to the hearts of the people he was trying to reach.

The results back in 1984 were astounding, and the program continues to amaze today. Yields in Farming God's Way plots double or triple the first year, and continue to increase year after year as the water holding capacity and fertility of the soil is restored.

This is how Craig Sorley, my colleague in Kenya who works with this program, explains it:

When the rainfall conditions remain near normal from year to year, the yields will increase year after year. When rainfall declines the Farming God's Way plot will also decline, but not nearly as much as the declines seen under conventional farming.

Also, during those first two to three growing seasons the primary yield benefits come from the water conservation features of FGW, and then, as time progresses, the benefits of restored soil structure and fertility kick in as well. Depending on the actual location, and where you started from, after roughly 5 or 6 years the plot is restored to its original potential. (Craig Sorley, Private correspondence with the author. For more information on Sorley's approach to agriculture, see his *Farming that Brings Glory to God and Hope for the Hungry*)

Care of Creation has been testing this program and teaching it to local farmers in central Kenya's Rift Valley region for a number of years, and even though our testing has coincided with several years of severe drought, our results confirm those of other practitioners.

Table 1:   Care of Creation Kenya.  Farming God's Way
Yield Results, 2010-2012. Project location: Kijabe, Kenya.

| Year | Crop Variety | Plot | Yield (kg/acre) | Yield Ratio | Comments |
|---|---|---|---|---|---|
| 2010/11 (winter/spring) | Maize | FGW | 484 | N/A | Drought year |
| | Maize | Traditional | 0 | | |
| | Beans | FGW | 234 | 370% | Variety: Rose C |
| | Beans | Traditional | 63.3 | | " |
| 2011 (summer) | Beans | FGW | 1412.4 | 132% | Normal rains. Variety: Dollycott |
| | Beans | Traditional | 1066.7 | | " |
| | Onions | FGW | 7802.8 | 460% | |
| | Onions | Traditional | 1707.7 | | |
| | Maize | FGW | 2370.5 | 120% | |
| | Maize | Traditional | 1976.4 | | |
| 2011/12 (winter/spring) | Beans | FGW | 1086.5 | 320% | Dry conditions: Variety: Dollycott |
| | Beans | Traditional | 335.8 | | " |
| | Maize | FGW | 1708.7 | 194% | |
| | Maize | Traditional | 879.0 | | |
| | Potatoes | FGW | 8790.5 | 175% | |
| | Potatoes | Traditional | 5037.3 | | |

Table 1 shows sample yield results from the first three years of testing. Increases of this magnitude are more significant than most of us realize. These numbers are the difference between just having enough to eat and being able feed your children as much as they want. They represent school fees and new shoes and medicine for people who have no other source of cash income than what they might be able to sell in the local market.

What is the take-away lesson? Things go better for people when we work with God and in harmony with his creation.

If we want to try to run the world differently than it has ever been done before, maybe we should try to figure out how to work with God and his creation in the same way. Maybe we could apply the principles from one small farming program to the entire human enterprise by asking these questions: What might it look like to run a factory God's way? Or a school? Would it be possible to run a country's political system God's way? How about the whole world? Can we run the world God's way?

**Been there, tried that?**

But we're back with the problem of history, again. I can hear your next objection:

People have tried this before, Ed. And it never works out.

True enough. John Calvin tried it in Geneva, Switzerland. The Roman Catholic Church tried it for centuries. Some of the first European settlers in North America tried it in Plymouth Colony in what is now Massachusetts. We could include Cromwell's England, Saudi Arabia, Iran after the revolution and the Taliban's Afghanistan on the list. All represent attempts to run a city, a colony or a country as the leaders thought God would do it. Almost all were or are marked by dictatorial rule, intolerance and state-sponsored violence toward dissenters. None were very happy societies, or very prosperous. Every attempt in history to create a society like this has been a failure.

Theocracies don't work. And a theocracy won't work as long as the underlying systems by which society operates stay the same.

Remember what we said several pages ago? Maybe the problem is that we haven't been doing things differently enough.

All of these historical attempts to do things "God's way" failed. There were many different reasons, and scholars have examined each of them exhaustively. But it seems to me that they all have one failing in common: Each one was a top-down effort that placed a religious overlay on the same human political and economic systems. A feudal system is still feudalism whether it is run by a feudal Lord, a Pope or an Ayotollah.

Imagine if Brian Oldrieve had tried to reform his farm by instituting staff prayer times, but kept on abusing the land, soil and water as had always been done before? Of course it would have been a failure. And that is what all of these attempts to "do things God's way" really were. They were all endeavors to put an imagined God-facade on a system that was fundamentally broken. No wonder nothing changed.

No, if we're going to do things differently, we have to start from scratch.

# 3

## What Does God Want?

If we want to run the world as God would have us run it, we need to figure out what God wants. What are his purposes? What outcomes is he looking for? What does God want from us and from creation? What are God's goals?

### Running God's ranch

Suppose you own a large ranch in, say, Texas, and you want me to manage it for you. The length of time is undetermined, but my tenure is not going to be permanent. You are off on a trip of some kind, you will be returning, but you're not sure when. So you give me complete legal authority over this property: You give me Power of Attorney, sign me on to your bank accounts, and off you go into the sunset, leaving me to...

To do what, exactly? What am I supposed to do with all of this?

Run the ranch, of course.

But how? To what ends?

As I sit at (your) desk, looking out over the property, I consider my options. You have given me complete control of everything. I could pursue short-term profit taking. I could sell off timber, cash in water rights, and hold livestock auctions. I might make a great deal of money for you in a very short time. When you came back, the bank accounts could be enormous, yet there might be nothing left of the ranch.

Alternatively, I could manage conservatively with a long-term horizon in mind. I might have studies done to determine what strategy of harvesting would maximize the health and viability of the woodlands while still providing cash income to run the ranch. I would want to learn how to manage grazing lands in a way that would keep them healthy. How many cattle could I keep in each quarter-section at a time without damaging the ground cover and causing erosion and long term decline?

How do I decide what to do?

I hope that early in our negotiations, you and I would have had an important conversation. You would have sat down with me and talked about what your own goals are for the property. Maybe we would have walked or ridden around the ranch, and you would have shared with me your dreams, the things you had in mind when you first bought the property. As your caretaker, your steward, I should want my management strategy to follow yours as closely as I can. I am only a manager. My task is not to come up with my own goals, but to develop strategies that would achieve your goals for you. I would want you to be pleased with my efforts when you finally returned.

Do you see?

Everything I do as your manager should be guided by your dreams and goals.

If you're planning to sell it in a year, maybe the short-term, cash-it-all-in approach is what you want from me. In that case you would be pleased to come back to an empty ranch and a brimming bank account. But it might be that you are looking at this as an investment that you want to pass down to your grandchildren someday. If that's the case, the last thing you want me to do is liquidate everything in the first year. You will want to be sure that I understand that long-term viability is your goal.

### God's goals, not ours

It's not hard at all to make the connection between your ranch and God's creation, is it?

This is God's place. God has hired us. His goals should determine how we do the job.

God has hired us? Has he? Really? Let's go back to the ranch analogy.

I took over management responsibilities because you asked me to do the job. As property owner, you can have anyone – or no one – run your ranch. And whether you've done a nationwide search for a manager with top qualifications, or you've picked me out of the cowhands in the bunkhouse doesn't matter. The point is that you, the owner, chose me. That's the only way I can even be on the property, let alone to make decisions that affect every aspect of the operation.

What makes us think we can assume the task of managing God's world? Do we have any right to be talking about running the world at all? Has God really "hired" us? If it's God's world, do we have any right to be in creation's executive offices at all?

In organizations and governments, an administrator or officer cannot make a decision or rule on an application for a permit unless they are "competent", in a legal sense, to do so. Competence does not refer to ability, but to authority. The Building Department may give you a permit to build a new house; the Department of Education may not do so; it is not competent in the area of building permits. But the DOE can give you a diploma of graduation from high school, something the Building Department cannot, no matter how much they might think you deserve it. The Building Department is not competent to issue diplomas. It has no authority. Similarly, I am competent to sign checks for my organization because of the authority that has been given to me by our Board of Directors. You are not. If you were to get hold of the company checkbook, you would be physically able to sign checks – but you would not be legally competent to do so. In fact, you would be committing a crime. Something I am required to do as a matter of routine would put you in jail.

The question before us is an important one: What is our standing as human beings with regard to God's creation? Are we competent to manage the day-to-day operations of our world? Can we sign the checks?

Some answer with a resounding "No!". We are no better than any other creature. To argue otherwise is speciesism, they say, discrimination against other species. The term was apparently first used by Richard Ryder, a British psychologist. "I use the word 'speciesism'," Ryder wrote, "to describe the widespread discrimination that is practised by man against other species." (Ryder 1983)

The same point of view is shared Richard Dawkins and many others. We are only animals who happen to be at the top of the heap right now through the accident of evolution. To suggest we have rights or authority toward or over anyone is presumptuous.

Others might agree that we have no right to assume authority over other creatures, but would argue that our higher intellect creates an obligation on our part to manage creation. Since we have a demonstrated ability to destroy other creatures and their habitats, we have a corresponding responsibility to not do that which we are capable of doing.

Still others might even say that the question itself is invalid: We are just the end product of a long process of natural development. Our intellectual abilities and what they produce are equally part of that evolutionary process, so it really doesn't matter what we do – what will be will be…

### God's opinion

We must take a different approach to this problem. We've labeled the world "God's creation". This presupposes both God's existence and his role as Creator, Owner and Lord of this world.

That premise is in turn based on another presupposition: We can learn about God and what he wants in the Bible. So maybe what we ought to do is ask God himself, through the Bible, about this question of competency and our role in his creation.

It doesn't take much digging. We find an answer in Genesis, the first book of the Bible, on the very first page:

> *Then God said, "Let us make mankind in our image, in our likeness, so that they may rule over the fish in the sea and the birds in the sky, over the livestock and all the wild animals, and over all the creatures that move along the ground."*
> *So God created mankind in his own image,*
> *in the image of God he created them;*
> *male and female he created them.*
> *God blessed them and said to them, "Be fruitful and increase in number; fill the earth and subdue it. Rule over the fish in the sea and the birds in the sky and over every living creature that moves on the ground."* Genesis 1:26-28

The same idea appears in other parts of the Bible. One of the clearest is in Psalm 8, a passage we will return to:

*You have given him dominion over the works of your hands;*
*you have put all things under his feet,*
*all sheep and oxen,*
 *and also the beasts of the field,*
 *the birds of the heavens, and the fish of the sea,*
 *whatever passes along the paths of the seas.*
Psalm 8:6-8

The answer to the question we began with in this chapter is a clear "Yes." We do have competency. We do have authority. God has given us the checkbook. We are in charge of the ranch.

**Now what?**

Let's go back to the ranch office: When we last visited it you explained to me your goals for your property so that I could make them mine. At this point the analogy breaks down.

We can't sit down with God in quite the same way that you and I might be able to. Our situation is as if you and I had never had that important conversation. As if you just sent me an email saying, "Sorry for the short notice. I have to leave tonight. Please come and take over the ranch. The keys are under the flower pot and the notes are in the office…"

When I arrive, I find the keys, the checkbook, a Power of Attorney document, and lots and lots of files. You've left reams of paper with thoughts on many things. But there is no memo headed "Goals for the Ranch". It looks like I'm going to have to work a bit to figure out what you intend for me to do.

This is exactly how it is with God and our role in creation. We have the Power of Attorney. God has clearly told us we're in charge. Genesis 1 and Psalm 8 can hardly be interpreted otherwise. And we have the entire Bible filled with God's thoughts about many things. But he has not told us in so many words what we're supposed to do now that we're sitting behind the desk.

We're going to have to do a bit of digging.

# 4

## DOMINION AND SOVEREIGNTY

This idea that human beings have been given authority over creation is often referred to as "dominion". This is based on the language of the first chapter of Genesis. Dominion is a topic I think about a lot; a great deal depends on how we interpret the concept of human authority over creation.

Mishandled, it can be, and has been, the excuse for all kinds of selfishness, arrogance and abuse of God's world. Understood correctly, it is the key to figuring out where we fit in God's creation and how to do the task God has given us to do – how to have a world that reflects God's goals for it.

One morning several years ago in the course of my regular Bible reading I found myself at Psalm 8, the biblical poem we quoted in the last chapter. It struck me that if there is a key to a proper understanding of dominion anywhere, it will be here. So I stayed with that Psalm for weeks – meditating on it constantly, reading various commentaries, and thinking, thinking, thinking.

You know how a song gets stuck in your head? That's what Psalm 8 did to me.

One morning everything came clear, just as when you finally get one piece of a jigsaw puzzle to fit, and five or ten others fall into place. That's what it felt like.

Here is the Psalm:

> *O Lord, our Lord,*
> *how majestic is your name in all the earth!*
> *You have set your glory above the heavens.*
> *Out of the mouth of babies and infants,*

*you have established strength because of your foes,*
  *to still the enemy and the avenger.*
*When I look at your heavens, the work of your fingers,*
  *the moon and the stars, which you have set in place,*
*what is man that you are mindful of him,*
  *and the son of man that you care for him?*
*Yet you have made him a little lower than the heavenly beings*
  *and crowned him with glory and honor.*
*You have given him dominion over the works of your hands;*
  *you have put all things under his feet,*
*all sheep and oxen,*
  *and also the beasts of the field,*
*the birds of the heavens, and the fish of the sea,*
  *whatever passes along the paths of the seas.*
*O Lord, our Lord,*
  *how majestic is your name in all the earth!*

## Divine bookends

I found the missing pieces of the puzzle at the beginning and end of the poem. These are identical, a pair of bookends, framing the rest of the Psalm:

*"O LORD our Lord, how majestic is your name in all the earth..."*

Whatever the psalm might say about human beings in the middle – and it says a lot! – it begins and ends with God. More specifically, it begins and ends with the majesty and beauty of God's name.

It's easy to miss the significance of the first phrase in English. If you are reading in French or many other languages, you can more easily get the meaning. In English the repetition of "Lord" sounds like poetic repetition. It isn't - there are actually two Hebrew words in play here. The first occurrence of LORD translates the Hebrew Yahweh, the name God gives to himself in the burning bush encounter with Moses:

God also said to Moses, "Say this to the people of Israel, 'The LORD

[YHWH, or Yahweh], the God of your fathers, the God of Abraham, the God of Isaac, and the God of Jacob, has sent me to you.' This is my name forever, and thus I am to be remembered throughout all generations. (Exodus 3:15)

That is exactly what happened. The God of Israel is known as Yahweh, LORD, from that time forward. There's a lot of theology wrapped up in this, but what is important for our purposes is not what the name means, but the simple, astounding fact that God wants Moses, and later David and all of his people, to know his name. Hang on to that. We're going to come back to it a number of times.

The second occurrence of "Lord" translates the Hebrew word Adonai. This is a functional term, a job title. It is used for anyone who has authority or control over someone else – a master over his slaves, for example, or a foreman on a construction crew. Here David acknowledges that his God, Yahweh, is in a position of authority over him.

The beginning and end of Psalm 8 together affirm God's control and sovereignty over us, his people:

*O Yahweh, you who are our absolute master and sovereign, how majestic is your name in all the earth...*

And the point is?

Just this: Our authority over God's creation begins and ends with God's authority over us.

We asked earlier whether we human beings have authority or dominion over God's world. We saw that this is the case both in Genesis 1 and in verses 6-8 of this psalm. What we are learning now is that our authority over God's creation is subordinate and subject to God's authority over us.

### A cosmic golden rule

Whatever human dominion over creation turns out to mean, it is book-ended – limited, constrained, guided, confined – by God's dominion over us. The actions, purposes and goals we pursue in our management of God's creation must reflect God's purposes for us. We should be pursuing God's goals for his creation and we can best do that by looking at the goals God has as he cares for us. It's like a creation-wide version of the golden rule: Not just that we should 'do unto others as we want them to do unto us' (paraphrased, see Matthew 7:12) but that we should care for all of creation as God cares for us.

Two things follow:

*First, an observation:* God's care for us is marked by tenderness, compassion and mercy. He is a loving father, a gentle guide, a forgiving savior. His approach to us is that of the prodigal son's father, waiting by the gate daily to welcome us home (Luke 15:11-32). It is that of the shepherd who searches long for a lost sheep and brings it home rejoicing. God's care for us is in every case almost the opposite of how we act toward his creation. God gives; we take. God seeks our best. Those God cares for blossom and flourish under his care. We, by contrast, seek from creation what is best for ourselves. Creation withers and dies under our hands.

*Second, a question:* If we apply this principle as a test, how much of the present human enterprise (society, economy, business, government, and everything else) will pass that test? With a society built on an economic system fueled by greed, where benefits accrue to those who destroy God's creation more than to those who seek to preserve it, how much of what we do as human beings comes even close to meeting such a standard? Not much.

If we want to be serious about running God's world God's way, we've probably got to start over.

What does that mean? You might be tempted to jump to a political or economic conclusion here – wondering what kind of a revolution I'm trying to start. Yes, there almost certainly are political and economic implications when we ask questions like this. But let's not get too far ahead of ourselves. Before we think about what kind of world we might want to have, we need to learn more about God's goals for us.

Let's go there.

# 5

## GOD'S GOALS

How does God want to care for us? What does he want to do for us?

This seems like an odd question. Shouldn't we be asking what we should be doing for God? Isn't the essence of religion figuring out what God wants from us? That is how the question is usually asked and answered. The opening of the Westminster Shorter Catechism is a case in point:

What is the chief end of man?

The chief end of man is to glorify God and enjoy him forever.

This is true. We do exist to serve God. Our "chief end", our ultimate purpose, is to "glorify God and enjoy him forever." It seems entirely appropriate that we should be asking what God wants from us, not the other way around.

But there's another side to this. In fact, God is the initiator in our relationship with him. He is the lover, we are the beloved:

> In this is love, not that we have loved God but that he loved us and sent his Son to be the propitiation for our sins. (I John 4:10)

Similarly, in Romans 8, the entire focus of our hope in God is not on what we might be able to do for God, but on his love for us:

> No, in all these things we are more than conquerors through him who loved us. For I am sure that neither death nor life, nor angels nor rulers, nor things present nor things to come, nor powers, nor height nor depth, nor anything else in all creation, will be able to separate us from the love of God in Christ Jesus our Lord. (Romans 8:37-39)

It is God's love for me that can never be overcome, not my love for God.

The best way to learn love is really quite easy: Let God love you. And then, learn from his love for you how to love other people. That is exactly how we want to learn how to rule God's creation, by understanding how God rules us. Where should we go to find out what God's goals are in his rule over us?

**"This is how you should pray..."**
One place to look might be the prayer that we commonly call the Lord's Prayer, in which Jesus taught his disciples how to pray.

> *Pray then like this:*
> *Our Father in heaven, hallowed be your name,*
> *Your kingdom come, your will be done,*
> *   on earth as it is in heaven.*
> *Give us this day our daily bread, and forgive us our debts,*
> *   as we also have forgiven our debtors.*
> *And lead us not into temptation,*
> *   but deliver us from evil.*
>                                             Matthew 6:9-13

Think about what is happening here. Jesus is God. Thus, this prayer is God telling us what to pray for. This is what God wants us to want. It's a summary of God's goals for us.

A number of years ago I was living in Pakistan and needed to apply for a visa extension. Granting the extension would require the authorities to make an exception for me. What I needed was not something that would normally be approved. I went to discuss my case with the officer in charge of visa applications. We had a good relationship. He was considerate and sympathetic. It was clear that he wanted to help me, but he also had rules he had to follow.

He told me what I needed to do. I would need to file an application with all of the proper forms filled out in duplicate (or triplicate, probably). I would also need to write a letter explaining why I should be granted this exception. To whom should I address the letter? To him, of course – he was the one who would approve it! And then he proceeded to tell me exactly what I should say in the letter I was going to bring to him.

Do you see? The officer who would approve my request was telling me how to state my request so that he could approve it. I wrote the letter, just as I was told to. And yes, my visa was granted. I have since found that the same strategy often works with fundraising. A potential donor will tell me what he wants me to ask him for; I'll put that in a letter to him, and usually with no surprise, my funding request is granted.

That is what is happening in the Lord's Prayer: God wants to answer our prayers, so he is telling us what to ask for, and how to ask, so that he can approve our requests.

Based on this prayer, then, what does God want for us?

**What is important to God?**

First, God wants his name to be proclaimed: "Hallowed be your name". "Hallowed" is an old term that captures a range of meaning. God wants his name to be revered, respected, known, proclaimed. God wants to be known as God.

Second, God wants his kingdom to be established. "Your kingdom come, your will be done." The parallel phrasing is important. God's kingdom "comes" when his will is done. What is reality in heaven is his goal on earth. We'll be exploring what this means in terms of a society that reflects God's 'kingdom values.'

And third, God wants to provide for people: "Give us this day our daily bread...forgive us our debts... lead us not into temptation." All of these requests have to do with human needs for physical and spiritual sustenance.

In *The Message* Eugene Peterson captures what we see in this prayer perfectly:

> *Our Father in heaven,*
> *Reveal who you are. (i.e. proclaim your name!)*
> *Set the world right; (establish your kingdom!)*
> *Do what's best— as above, so below.*
> *Keep us alive with three square meals. (provide for your people!)*

*Keep us forgiven with you and forgiving others.*
*Keep us safe from ourselves and the Devil.*
                    Mt 6:9-13 *The Message.* Emphasis added.


**A quick review...**

This three-part formula – name, kingdom, people – is the heart of what we're trying to understand in this study. We're going to go on and develop each of these, for they are rich and fascinating topics. But first let's pause and see how we got here:

We established that *God has placed us in his creation with real authority or dominion.* We are like managers on a ranch. We have the keys and the checkbook.

This means we have freedom and authority to do pretty much what we want with God's creation; but to be successful in God's eyes, *our goals in managing God's creation will have to be consistent with his goals.*

We found in Psalm 8 that our rule or dominion over God's creation is bookended by God's rule over us. Thus we concluded that *how God governs us and his goals for us can be a model as to how we ought to be governing creation.*

And in Jesus' teaching on prayer we have seen a way of discovering what God cares about most, since what he cares about will be reflected in his goals for us and creation. This led us to the conclusion that God cares about his name, his kingdom, and his people.

Back to Psalm 8

In Psalm 8 we find the same three themes – name, kingdom, people. Figure 1 shows how I marked up the Psalm when I first did this study:

If we really want a world that reflects God's goals, these three principles should be able to point the way. We'll explore these in detail in the next part of our study.

Figure 1:  Psalm 8

Name
> ¹ *Lord, our Lord,*
>     *how majestic is your name in all the earth!*
> *You have set your glory*
>     *in the heavens.*

Kingdom
> ² *Through the praise of children and infants*
>     *you have established a stronghold against your enemies,*
>     *to silence the foe and the avenger.*

People
> ³ *When I consider your heavens,*
>     *the work of your fingers,*
> *the moon and the stars,*
>     *which you have set in place,*
> ⁴ *what is mankind that you are mindful of them,*
>     *human beings that you care for them?*
> ⁵ *You have made them a little lower than the angels*
>     *and crowned them with glory and honor.*
> ⁶ *You made them rulers over the works of your hands;*
>     *you put everything under their feet:*
> ⁷ *all flocks and herds,*
>     *and the animals of the wild,*
> ⁸ *the birds in the sky,*
>     *and the fish in the sea,*
>     *all that swim the paths of the seas.*

Name
> ⁹ *Lord, our Lord,*
>     *how majestic is your name in all the earth!*
>     *how majestic is your name in all the earth!*

# Part II
# God's World, God's Goals

IN WHICH WE EXPAND ON THREE GOALS THAT SEEM
IMPORTANT TO GOD: TO PROCLAIM HIS NAME, TO
ESTABLISH HIS KINGDOM, AND TO CARE FOR HIS
PEOPLE.

## Goal One: Proclaim His Name

'Hallowed be your name'

'O Lord, our Lord, how majestic is your name in all the earth. You have set your glory above the heavens.'

# 6

## A Name is More than Just a Name

In the ancient world, a name was more than a word, a sound, letters on a page. It was not just a label. It reflected the essence, the very nature or being of the person or thing being named.

For Shakespeare, "that which we call a rose by any other name would smell as sweet" [*Romeo and Juliet* II ii 1-2] but ancient Greeks or Hebrews would have disagreed. If Shakespeare was right, they might have said, then the rose didn't have the right name in the first place. The name that an object (or a person) has matters, because it reflects the true essence of the thing or person.

When God asked Adam to name the animals (Genesis 2:19) something important was going on. God was saying to Adam, "Understand the creatures; study them; give them names that reflect the essence of their being" with the result that "whatever the man called each living creature, that was its name."

**Who are you, God?**

This background explains why it was such a big deal when Moses dared to ask God his name in the passage we looked at earlier:

> "If I come to the people of Israel and say to them, 'The God of your fathers has sent me to you,' and they ask me, 'What is his name?' what shall I say to them?" (Exodus 3:13)

Moses intrudes on God's personal space in a big way. Even more surprising, God took Moses' question seriously and answered him:

> God said to Moses, "I Am Who I Am." And he said, "Say this to the people of Israel, 'I Am [Yahweh] has sent me to you.'" God also said to

Moses, "Say this to the people of Israel, 'The LORD [YAHWEH], the God of your fathers, the God of Abraham, the God of Isaac, and the God of Jacob, has sent me to you.' This is my name forever, and thus I am to be remembered throughout all generations. (Exodus 3:14-15)

What we are concerned with here is not what God called himself, as interesting and important as that is. What is striking is that God told Moses what he wanted to be called. God wants his name to be known among human beings. God wants us to know his essence, his being. God wants to be known.

Similarly, in the Lord's prayer, Jesus tells us that we should pray that God's name will "be hallowed". What does that mean? The word means to make something holy, to set it apart, to give it respect and reverence. How do we hallow God's name? By knowing it. By knowing him in the sense used in the ancient world. By giving him the place of respect he deserves in our own lives.

In Psalm 8 David brings another dimension into the discussion. God's name is majestic and his glory is above the heavens. One of the ways in which God makes his name, that is himself, known is through creation.

God's self-revelation in creation is a major theme in the Bible. "The whole earth is full of his glory!" proclaim the Seraphim in Isaiah 6, and God's "invisible attributes," says Paul,

"namely, his eternal power and divine nature, have been clearly perceived, ever since the creation of the world, in the things that have been made." (Romans 1:19-20)

### Broadcasting in stereo

This is not to suggest that God has not also made his name known through the Bible. Of course he has. It should be obvious: The Lord's prayer assumes it, and our evidence for the fact that God makes his name known outside of the Bible comes from the Bible itself. The Bible does not exclude itself when it shows us another way in which God makes his name known.

But does it matter? Perhaps knowing that God's name is proclaimed in creation is only important for people who don't have the Bible? For those of us who have the 'better' revelation of the Bible, perhaps God's name in creation is redundant or at least unimportant.

Not so.

Here is another way to understand this. God is broadcasting on two channels at the same time, through His Word and His World. Revelation coming to us in stereo.

This idea is very clear in Psalm 19, which begins much like Psalm 8:

> *The heavens declare the glory of God,*
> *and the sky above proclaims his handiwork.*
> *Day to day pours out speech,*
> *and night to night reveals knowledge.*
> *There is no speech, nor are there words,*
> *whose voice is not heard.*
> *Their voice goes out through all the earth,*
> *and their words to the end of the world.*

But look where the psalmist goes in the second half of the Psalm:

> *The law of the LORD is perfect,*
> *reviving the soul;*
> *the testimony of the LORD is sure,*
> *making wise the simple;*
> *the precepts of the LORD are right,*
> *rejoicing the heart;*
> *the commandment of the LORD is pure,*
> *enlightening the eyes;*
> *the fear of the LORD is clean,*
> *enduring forever;*
> *the rules of the LORD are true,*
> *and righteous altogether.*
> *More to be desired are they than gold,*
> *even much fine gold;*
> *sweeter also than honey*
> *and drippings of the honeycomb.*

The first half is about God's glory, his self-proclamation, wordlessly, through creation. The heavens, the sky, day and night – all of these proclaim his handiwork, pour out speech, and reveal knowledge – without speech and without words. The rest of the psalm is about the written word: Law, testimony, precepts, commandments, rules are all synonyms for the vehicle of revelation that we know as the Bible.

Two channels: God's World and God's Word, the Book of Creation and the Book of Revelation. Hang on to this idea; when we start talking about how to translate God's desire to make himself known into our goals for his creation, these two ways that God proclaims his name will turn out to be immensely important.

### Video, audio or both?

Maybe "stereo" isn't the most adequate analogy. Let's think of it as if God were sending his message to us through a device like a television that has both video and audio channels. The created world is the video; the Bible is the audio.

Both video and audio carry information, but there is a lot more information on the audio channel than on video, and it tends to be more accurate information. Take any typical television show or movie. Watch for a while with the mute button on. You'll get some idea of what's going on, but you're going to miss a lot. You will also be forced to do a lot of guessing and interpretation, and you might well get it wrong. You thought they were fighting over the girl, but they were really fighting over who pays the bill.

Now do the opposite. Try listening to the program with the picture blocked. The experience is not complete, and not nearly as rich, but it is much easier to follow the story that way than the other. There is more content on the audio channel than on the video.

Most of the content being broadcast is carried on audio, but that doesn't mean video isn't important! Having pictures enriches the entire experience, or we wouldn't be spending so much on larger and larger screens every year. And there are some cases – sports events come to mind – where video is essential. The Super Bowl without a picture would not be much fun.

Yes, we can learn about God through the Bible alone (like a television with the picture turned off). And it is possible to learn a great deal about God through creation apart from the Bible (remember Romans 1:19). Neither one is a complete experience without the other. God is proclaiming his name through both creation and through the Bible, and he wants and expects us to have the full, rich experience of both channels.

If our priorities are to be God's priorities, our first goal will be to have a world in which God's name is proclaimed. We will want to proclaim his name just as he does, through his Word, and in his World.

What might that look like?

# 7

## Proclaiming His Name Through Witness and Worship

The fact that God wants to proclaim his name is clear throughout the history of the human race as told in the Bible. It is a story of God pursuing people by revealing himself to them.

### A love story from the very beginning

It starts at the beginning, right after Creation. Adam's sin of disobedience in the Garden of Eden, recounted in Genesis 3, comes to light because God comes looking for Adam, not the other way around:

> And they heard the sound of the LORD God walking in the garden in the cool of the day, and the man and his wife hid themselves from the presence of the LORD God among the trees of the garden. But the LORD God called to the man and said to him, "Where are you?" (Genesis 3:8-9)

We see the same story repeated over and over again in the history of Israel. God chose Abraham and promised to make him into a great people (Genesis 15); he repeated the promise on behalf of Isaac (Genesis 17:19ff). He chose Jacob over Esau (Genesis 25:23), called Moses in the desert to lead the people out of Egypt to the Promised Land (Exodus 3), found David tending sheep and made him Israel's greatest King, calling him 'a man after his own heart' (I Samuel 13:14 & 16:1-13), made himself known to Solomon in a dream (I Kings 9).

In every case, we see God doing the pursuing, and the purpose of that pursuit is always so that God can make himself known to the person he is after.

He makes himself known with words: He proclaims his name. If this is one of God's goals – articulated in the Lord's Prayer, in Psalm 8 and throughout history – it needs to be our goal too.

### "Verbal" proclamation

Let's call this "verbal proclamation". This is the proclamation of his name through understanding, communicating and responding to his written word, the Bible.

At least three different "actions" are implicit in this concept:

Hearing God's name. Because we are born sinners, separated from God, we have to start on the receiving end of the process. We must have God's name proclaimed to us, usually by someone else, though sometimes this happens through a direct encounter with God through the Bible. This hearing God's name allows us to begin an actual relationship with God, something we usually think of as conversion or being born again. (John 3:1-21)

Speaking God's name back to him through prayer, worship and other means – spiritual disciplines – that Christians have developed over many centuries. This communication is personal and private, as well as communal and public. Church life has historically been centered on encouraging and practicing this kind of proclamation.

Sharing God's name with others who do not know him so they can be included in both the relationship and worship aspects. Some traditions call this witness, others evangelism, from the Greek evangel which means gospel or more directly, good news. This activity completes the circle. Having heard God's name from someone else, having learned to speak God's name back to him in prayer and worship, we then share his name with other people who have not yet heard, so they can be blessed as well.

All of these traditional Christian activities are different dimensions of the same goal, God's first priority. He wants to make his name known. When we are doing any of these things – experiencing a relationship with God, developing that relationship through worship, prayer, fellowship, and seeking to bring others into the circle of those who know God – we are beginning the process of shaping a world that looks the way God wants it to look, a world that reflects his goals rather than our own.

We can anchor this by stating it as the first of a number of principles that describe what the world would look like if it were to really be the world God wants. Here is our first:

**Principle #1: In a world that reflects God's goals, God's people will be energetically and effectively making God's name known personally and corporately, with activities directed back to God (prayer and worship) and directed to those who do not yet believe (evangelism).**

This objective is what most of us who are followers of Jesus understand to be the mission of the church. And we are doing it! Worship is happening. Prayer is taking place. People are witnessing to their faith and the church is growing in many parts of the world. While there are huge numbers of people who are not (yet) part of this, the verbal proclamation of God's name in our world today is remarkable. There is no country on earth that does not have someone who would call themselves a follower of Jesus. There are few if any cities of any size where you could not find some evidence that God's name is being proclaimed in this way.

But this isn't the whole picture. Remember our audio/video analogy? Verbal proclamation is critical, but it is still only one channel. We might congratulate ourselves that we have done a reasonably good job of proclaiming God's name verbally. Our audio is working fine. But what about the video? What about our proclamation of God's name through creation? We are about to see that this is important, too. If you are a broadcaster, the big game is about to start and your viewers can hear the announcers but have no picture, you haven't done your job.

That leads us to the second "channel" of God's proclamation of his name.

# 8

## Proclaiming His Name
## through a Flourishing Creation

We saw in chapter 7 that God intends that his name be proclaimed through two channels. We spoke of the Word and the World before. If God uses both, we have to use both. Our proclamation of his name through just one of those channels, even if it is more or less successful, is not sufficient.

But what does proclaiming his name through the world mean? What does it look like?

### "Biospherical" proclamation

I would like to call this biospherical proclamation – the proclamation of God's name through the care and nurturing of his creation, in particular the biosphere, the realm of living organisms. *The biosphere* is a term used frequently among biologists and ecologists to describe the world of living creatures. Though there is plenty of wonder and beauty in the non-living world, it is the world of DNA-based life that is of interest to these scientists, and as you will see in a moment, is of particular interest for us as well.

This is new territory for some of us, so I want to explain myself carefully. If verbal proclamation is the communication of God's name by use of human language, biospherical proclamation would be the communication of that name through the vocabulary of the created world, and in particular the living created world. Where that world flourishes, God's name is being loudly proclaimed. When it suffers or declines, the sound of God's name fades. In this sense, every gardener is a proclaimer of God's name, whether he or she knows it or not. And every person walking in a forest is immersed in a flood of praise and worship that the most fantastic human choir cannot hope to match.

Our task is to see that it continues. Biospherical proclamation requires the preservation and flourishing of creation as a central part of the proclamation task.

This term may be new, but the concept is not. We have already seen several of many biblical references that show that God reveals himself in creation. Church leaders have understood and taught these principles since the earliest years of the church. But it was never much of an issue. The flourishing of creation was taken for granted – it was human survival that was of greatest concern.

Our ancestors, prior to the industrial revolution, lived in relatively small human communities surrounded by wild, untamed nature. Their largest cities would have been small towns for us. Travel, whether by land or sea, was dangerous. In the struggle for survival, human beings lost as often as they won. When they looked to creation for revelation, they experienced humility as much as wonder, for in their day creation revealed human insignificance as well as God's power.

**We don't know our own strength**

Things are different now. Human "progress" has impacted the rest of creation in ways that could not have been imagined by our ancestors. As *The Economist* describes it,

To think that the workings of so vast an entity could be lastingly changed by a species that has been scampering across its surface for less than 1% of its history seems, on the face of it, absurd. But it is not. Humans have become a force of nature reshaping the planet on a geological scale—but at a far-faster-than-geological speed. (*The Economist*, May 26 2011)

Through most of history, seeing God's name proclaimed in creation was a passive concept. It required no effort. Creation was just there. God's name, as in Psalm 8 and Psalm 19, was written across the skies for all to see, and evident in every tree and flower and the great beasts described in the latter chapters of the book of Job.

Not so much today. We have largely hidden ourselves from that revelation. The stars are still there, but we have to make an effort to see them. We're not often outside at night, and when we are, streetlights and shopping malls obscure our vision. There are still mountains and sunsets, but we spend our days in cocoons of artificial light and artificial air, staring at computer screens. When we do glimpse nature, even in worship, it is probably in the form of electronic pixels. When we venture outside and open our eyes, we more often see smog and streetlights than the splendor of God's name splashed across the heavens.

And worse, we are actively destroying large parts of that revelation. Rather than proclaiming God's name in the biosphere, we are busy erasing it. Fields become parking lots. Mountains are reduced to rubble to get at the coal underneath them. Coral reefs are bleached and dying. The list of endangered and extinct species, every one of which proclaims God's name, grows longer and longer.

Can we really say that we are proclaiming God's name in creation when the things we build and the way we live diminishes and even destroys it?

## The language of life

We can't plead ignorance. We know more about God's creation than human beings ever have. This knowledge can and should lead us to a deeper and richer appreciation of what God's creation really is, but it also leaves us without excuse for our actions.

Consider DNA – deoxyribonucleic acid. In less than sixty years we have moved from a vague understanding of the role of DNA to almost routine mapping of the entire genomes of organisms. We've learned that DNA is a kind of a "blueprint for life", and we have learned how it works. DNA is a language. It has letters ("bases"), words, sentences and paragraphs (genes). DNA even has a punctuation mark. DNA is a language that describes life.

Listen to Francis Collins, the head of the human genome project, describe the human genome. He sees it as a book "written in the DNA language by which God spoke life into being." He goes on:

I felt an overwhelming sense of awe in surveying this most significant of all biological texts. Yes, it is written in a language we understand very poorly, and it will take decades, if not centuries, to understand its instructions, but we had crossed a one-way bridge into profoundly new territory. (Collins 126-7)

If DNA is in fact "the language by which God spoke life into being", then the biological world is more than a backdrop for God's interactions with human beings. This biosphere is the words of God made visible.

The implications are profound.

## Two channels, one message

God's two channels of revelation – his Word and his World – are more closely related than we thought. Both are a result of God's desire to speak. In the first instance, he spoke words that gave us a message about himself; in the second, but historically prior instance, he spoke a living world into being. Earlier we talked about God proclaiming his name in the Word and the World. Now we need to modify that slightly, saying that God proclaims his name through the Written Word and the Created Word.

These two ideas, Written Word and Created Word, are tied together through another Word concept in the Bible.

In the beginning was the Word, and the Word was with God, and the Word was God; all things were made by him. (John 1:1)

This verse is of course speaking of Jesus. He is the Living Word, and as such is the ultimate fulfillment of the Written Word:

In the past, God spoke to our forefathers through the prophets (e.g. the written word) at many times and in various ways, but in these last days he has spoken to us through his Son (Jesus)... (Hebrews 1:1-2)

We also know that this Living Word is the creator and sustainer of the Created Word, the world:

He is the image of the invisible God, the firstborn over all creation. For by him all things were created...in him all things hold together. (Colossians 1:15)

The great hope of modern physics is the discovery of a grand "theory of everything". Here we have a divine theory of everything:

In him we have redemption through his blood, the forgiveness of our trespasses, according to the riches of his grace, which he lavished upon us, in all wisdom and insight making known to us the mystery of his will, according to his purpose, which he set forth in Christ as a plan for the fullness of time, to unite all things in him, things in heaven and things on earth. (Ephesians 1:7-10)

We have just opened the door to a massive topic that we don't have time to do justice to here. Entire volumes have been written on the person of Jesus Christ. For purposes of our study, though, this is what we need to understand: While God has indeed revealed himself to us through these three separate methods – the created word, the written word, and his son, Jesus – these means of revelation are not equal. The created word does indeed reveal God faithfully, though not completely. We need the detailed explanations of the written word to help us to understand and to come into a relationship with the God of creation. But even the written word, as precious and wonderful as it is, does not stand on its own – its true value lies in the fact that, as we noted in the verses from Hebrews 1 above, it points to the ultimate revelation of God in his son, Jesus Christ. No wonder he is called Immanuel – "God With Us". That is what he is.

**Why this matters**

If we are intending to shape the world so that it reflects God's goals, we need to have a society in which God's name is proclaimed in the same ways that he himself proclaims it. Using the terms we defined, above, we need to pursue both verbal and biospherical proclamation.

Based on all of this, let's add another principle:

**Principle #2: In a world that reflects God's goals, all human activity will proclaim God's name by working in harmony with, and contributing to the flourishing of non-human creation.**

If we were to measure present human activities against this objective, we would have to conclude that today's world is a failure. We covered this ground in Chapter 1, and there is no need to cover it again; what we need to understand is that the general degradation of God's creation isn't something we can ignore. It isn't just "how the world is right now"; this is disobedience and an indictment against us.

This failure is not something on paper; it has important, immediate consequences. We began with an analogy from the Farming God's Way program: The idea that if we do things God's way, they'll work out better for us. The obvious corollary is that if we don't do things God's way, we're going to have big problems. And we do.

Let's face it: The world is not working very well right now. We have an economy that doesn't work. A paralyzed political system. Environmental nightmares. Corruption and crime on every level. We have not been doing things God's way, and the results are obvious.

But can we really claim that if we were to turn things around by proclaiming God's name in these two ways, verbally and biospherically, all of our problems would be solved?

Yes and no. This is only the starting point, but it is a necessary one. Creating a human society is like building a house. You start with the foundation. Putting in a foundation doesn't give you a kitchen and a cozy bedroom – but you can't have those without it.

In my earlier book, *Our Father's World*, I talked about four key relationships – God, self, others and creation. These relationships have been broken by our sin, and are restored by God through Jesus Christ. The two kinds of proclamation we have been talking about lead to the restoration of two of these relationships.

Verbal proclamation, has to do with our relationship with God himself. If we want to try to manage God's world God's way, this is where we have to begin. To put it another way, it is impossible to have a world that reflects God's goals without involving God.

Biospherical proclamation speaks to a restored relationship with creation. This is a theological truth but it resonates with everything we know about biology, ecology and environmental science. It is simple fact that we are biological creatures. If God's non-human creation is not healthy and fruitful, we cannot long survive. Trying to manage God's world without caring for the rest of his creation is doomed to failure just as much as trying to do so without God.

Who wouldn't take a deal like this?

We ought to proclaim God's name because he wants us to. We need no other reason. Even if we had to sacrifice ourselves to please God, we should still obey. That is what it means to serve God. But we know from science that what we should have done out of obedience, for his sake alone, is exactly what we have to do to solve our own problems. Who wouldn't take a deal like this? Give God his due, care for his creation, all will be well with us. It is a classic win-win. And it raises a serious question. Why have we not obeyed?

Such disobedience is illogical and self-defeating. But there is a good reason why we live this way: If we were to proclaim God's name sincerely, we would be affirming him as Lord and King:

O LORD, our Lord, how majestic is your name in all the earth...

To name God Lord and King demands that we let him be our King. But we (human beings) have not wanted God as our King. There is a cosmic struggle for mastery going on. We can only have one King. If God is King, no one else can be.

This leads us to the second goal, the establishment of God's kingdom.

## Goal Two: Establish His Kingdom

May your Kingdom come, and your will be done, on earth as it is in heaven"

"From the lips of children and infants you have ordained praise, because of your enemies, to silence the foe and the avenger." (Psalm 8)

# 9

## GOD CARES ABOUT HIS KINGDOM

The "kingdom of God" – God establishing his authority on this earth, and by extension making things right again – is a major biblical theme. Like other topics we are touching on in this study, such as God's name or, still to come, the image of God in human beings, we don't have time to give the subject of God's kingdom the attention it deserves.

We can begin by noting that this was the touchstone of Jesus' own ministry. From the very beginning, Jesus indicated that his task was to announce the arrival of the kingdom of God or kingdom of heaven:

> From that time on, Jesus began to preach, "Repent, for the kingdom of heaven is at hand." (Matthew 4:17)

> These twelve Jesus sent out, instructing them, "Go nowhere among the Gentiles and enter no town of the Samaritans, but go rather to the lost sheep of the house of Israel. And proclaim as you go, saying, 'The kingdom of heaven is at hand.' (Matthew 10:5-7)

> Now after John was arrested, Jesus came into Galilee, proclaiming the gospel of God, and saying, "The time is fulfilled, and the kingdom of God is at hand; repent and believe in the gospel." (Mark 1:14-15)

> …and the people sought him and came to him, and would have kept him from leaving them, but he said to them, "I must preach the good news of the kingdom of God to the other towns as well; for I was sent for this purpose." (Luke 4:42-43)

And this wasn't just Jesus. John the Baptist prepared for Jesus arrival with the same theme: "Repent! The kingdom of heaven is at hand!" (Matthew 3:1). Both Jesus and John were announcing something that resonated with their Jewish audiences: The time had finally come when God would establish his kingdom – here, on earth, now.

This 'kingdom teaching' struck a chord. This was something the Jewish people had longed and dreamed about for centuries. It is no wonder that Jesus attracted crowds that numbered in the thousands. God was finally going to keep his promises. The kingdom would come. In fact, if you listened to Jesus carefully, you could conclude, as many did, that the kingdom had come and was already here.

**Already, not yet**

Which would explain why, Luke tells us, Jesus had to work to change the disciple's perspective:

> As they heard these things, he proceeded to tell a parable, because he was near to Jerusalem, and because they supposed that the kingdom of God was to appear immediately. (Luke 19:11)

The kingdom had, in fact, come with Jesus. It really was 'at hand' and those who believed were already living in the kingdom as Jesus indicated. But it had not yet 'appeared' in the sense that Jewish patriots longed for. God was not ready to wipe out the Roman empire; in fact, in a turn of events no one could have foreseen or understood, the kingdom's greatest victory was to be through the execution of its leader at the hands of those same Roman authorities. In a glorious swirl of theological perspective, it was all true: The kingdom *had come* in the person of Jesus. The kingdom *would be ultimately victorious* through defeat on the cross. And the kingdom *was still to come* when Jesus would be revealed in glory and all peoples on earth would bow before him.

We need to bring these perspectives to how Jesus touches on the kingdom of God in his prayer: He wants us to pray that the Kingdom will come. That is an implicit acknowledgement that while in some senses it is here, in other ways it has not come yet.

Even at the end, after his death and resurrection, the disciples asked, "Is it now? Will you at this time restore the Kingdom?" They still had in mind the grand, geographical and political fulfillment of the kingdom that had been the dream of the Jewish people for generations. Jesus' answer is directed to that vision, but also applies to the moral and ethical dimensions of the kingdom as well: Not quite yet. "It is not for you to know the times and the seasons..." (see Acts 1:6-7)

In spite of the 'Kingdom talk' that permeated Jesus' mission on earth, he came to the end still saying it wasn't yet time. The Kingdom had come, is still coming – and one day will be fully here, but this will be accomplished according to God's own timetable. The King will finally come when the King is ready to come.

## A cosmic shouting match

There is another element in the kingdom's delay that is important. Psalm 8:2 suggests what is causing that delay – the King has enemies.

*"From the lips of children and infants you have ordained praise,*
*because of your enemies, to silence the foe and the avenger."*
Psalm 8:2 NIV

This is a fascinating little verse. On the one hand, it hardly seems to fit with the rest of the psalm. Soaring praise to God is interrupted with thoughts of enemies, foes and avengers. But on the other hand the language fits the rest of the poem well, for this struggle with God's enemies is described in musical language.

God has enemies that need to be silenced. In the metaphor being used, these enemies are trying to drown out the sound of praise, the proclamation of God's name. And notice how God is going to rout his enemies. Not with sword, not with fire from heaven, but with praise from "the lips of children and infants." This conflict between good and evil is a cosmic shouting match, and God is going to win – but not by shouting louder. No, he will silence them with the weakest and most insignificant instruments he has.

Jesus saw this verse being fulfilled by his triumphal entry into Jerusalem:

> But when the chief priests and the scribes saw the wonderful things
> that he did, and the children crying out in the temple, "Hosanna to
> the Son of David!" they were indignant, and they said to him, "Do you
> hear what these are saying?" And Jesus said to them, "Yes; have you
> never read, "'Out of the mouth of infants and nursing babies you have
> prepared praise'?" (Matthew 21:15-16)

God's enemies throughout history have tried to drown out the sound of
his name on all of the channels he uses. Sometimes the opposition has
come through deception, words and ideas that contradict God's truth. Remember the serpent in the Genesis 3? "Has God really said...?" Our world
today is a cacophony of ideas screaming opposition to God and the proclamation of his name. Often opposition has taken the form of physical violence, as when followers of Jesus are persecuted, beaten and killed. This too
is happening today in many parts of the world. And a great deal of opposition to God in our world today comes almost invisibly in the form of materialism and wealth. Our possessions shout so loudly that it is difficult and
sometimes impossible to hear the name of God through the static. We are
surrounded by electronic toys and tools, burdened with caring for (and
paying for) cars and homes and clothing, and busy with activities that
crowd God's name out of our lives as effectively as any official rule of cen-
sorship might do.

Maybe this is what Jesus had in mind when he said, "it is easier for a
camel to go through the eye of a needle than for a rich person to enter the
kingdom of God." (Luke 18:25) And again when explaining a parable:

> And as for what fell among the thorns, they are those who hear, but
> as they go on their way they are choked by the cares and riches and
> pleasures of life, and their fruit does not mature. (Luke 8:14)

But there is another way that God's name is being silenced in our world
today: Through the systematic destruction of his creation. They – no, let's
be honest and say "we" this time – we are drowning God's voice in creation
through cruelty to his creatures, destruction and defacement of the beauti-

ful places he has made, and through rampant pollution. This is hard to hear! We are children of God; some of us have given our lives to the verbal proclamation of his name, yet by the way we live we are guilty of a crime of cosmic proportions: We have joined with God's enemies to drown out his name in creation.

Do you begin to understand why I am involved in creation care as a ministry? It isn't because I love plants more than people! It's because I understand now that creation is a three-dimensional representation of God's name, and the destruction of God's creation is an assault on God's name and, by extension, on his kingdom. Working to establish that kingdom is essential to defeating the enemies that would silence his name by destroying his world. Creation care, or Christian environmental stewardship, cannot be pursued within the confines of "environment" or "ecology" alone.

## Singing a heavenly song here on earth

So how do we do it? How do you and I work with God as he establishes his kingdom? What does this look like for us? How do we make the song of God's name heard in the lives we live here? The key to the coming of the Kingdom is found in the second part of the relevant phrase of the Lord's prayer: "Your Kingdom come, your will be done." God's Kingdom comes when God's will is done. The Kingdom is wherever God is being obeyed.

In this case, a negative illustrates the positive. Adam and Eve had to leave the Garden because they could no longer say 'Your will be done'. They had disobeyed the only command God had given them. Thus there was no place for them. They had to leave both Garden and Kingdom, because "Your will be done" was the one – and only – condition for living there.

Milton develops the same theme when he has Satan turning banishment into boast in one of the most famous lines in Paradise Lost:

> ...*Here at least*
> *We shall be free; the Almighty hath not built*
> *Here for his envy, will not drive us hence:*
> *Here we may reign secure, and in my choice*

*To reign is worth ambition though in Hell:*
*Better to reign in Hell, than serve in Heaven.*
                              Paradise Lost, Book I, lines 258-263

   To have a world that looks like God wants it to look, we need to see God's Kingdom established. And that happens when we do his will. Sounds easy enough. You think?

# 10

## The Greatest Movement in History

We've just concluded that God's Kingdom will be established as his people obey him.

But how? We usually think about this on the level of personal obedience to God's commands, and that is certainly where we have to begin. The Ten Commandments. The Sermon on the Mount. The two "Great Commandments" - love God, love your neighbor.

All are easy to understand, if not necessarily easy to accomplish, as guiding principles for us in our individual lives.

We should begin there – but my interest is on a different level. We have been exploring how we can reshape the world, and I want to ask how we can apply this concept to an entire society. Our problem is that if we use the vocabulary of "obedience", we open the door for the kind of dictatorial, top-down government-enforced "righteousness" that has been the mark of all the failed attempts of the past.

Thus a useful way to approach this would be in terms of values rather than laws. This is language we can apply to society as a whole without falling into a trap of legalism or politics. We might say it like this: If we are going to have a world that reflects God's goals, we need to develop a society – communities, economies, governments, everything – that functions with values that God would approve of, that is, kingdom values.

What would these values consist of? Or to put it another way, how would we know when this prayer – "your will be done" – has been answered?

### The most important commandments

For the Jews of Jesus' day, God's will was defined in the language of laws and commandments that we mentioned above. You are familiar with the Ten Commandments outlined in Exodus 20, of course, but there were thousands of others. Some were from the Bible, others were added through the centuries as interpretations. It was overwhelming. Where would a devout person even start?

Against this background, someone once asked Jesus, "Of all these commandments, which is most important?" It's the same question we're asking. Where should we begin?

All of the thousands of commandments that are trying to define "God's will" can be boiled down to just two, says Jesus:

> *And he said to him, "You shall love the Lord your God with all your heart and with all your soul and with all your mind. This is the great and first commandment. And a second is like it: You shall love your neighbor as yourself. On these two commandments depend all the Law and the Prophets."*
>
> Matthew 22:37-40

Love God. Love people. Loving God takes us back to what we've already learned about proclaiming his name. But what about loving people? What does that look like?

One of Israel's prophets, Micah, preaching centuries before Jesus, summarized God's will in similar language:

> *He has told you, O man, what is good;*
> *and what does the LORD require of you*
> *but to do justice, and to love kindness,*
> *and to walk humbly with your God?*
>
> Micah 6:8

Jesus' disciple John ties loving people to the example Jesus gave us when he died for us, and he applies this in a very practical way:

By this we know love, that he [Jesus] laid down his life for us, and we ought to lay down our lives for the brothers. But if anyone has the world's goods and sees his brother in need, yet closes his heart against him, how does God's love abide in him? Little children, let us not love in word or talk but in deed and in truth. (I John 3:16-18)

We now have content for what we started off by calling Kingdom values: Establishing God's Kingdom means developing a society marked by values like love, mercy and justice. And now we are ready to state a third objective for our new world:

**Principle #3: In a world that reflects God's goals, every dimension of society – political, economic, judicial, social, etc. – will operate according to kingdom values like love, mercy and justice.**

### Something everyone wants, but no one knows how to get

In late 2011, a new movement quite suddenly captured the attention of people around the world: "Occupy Wall Street" thrust itself into the public eye in the US and in dozens of other countries as well. No one knew where it had come from. And it turned out few knew where it was going. No one was in charge, and no one in the movement seemed to know what they wanted. The movement would not, or could not, articulate specific goals. Protester's complaints were expressed in the broadest terms: The economy isn't working. There is too much inequality. We need a world marked by justice, not greed.

Occupy Wall Street was pleading for a world marked by what we are calling kingdom values: Justice, love, mercy, kindness. But they didn't know how to get there.

I do a lot of work with people in the secular environmental movement. More and more I hear and read pleas for a new kind of society. People want a world that operates on different values than the world we now see. There is a growing consensus among environmentalists that much of the environmental crisis is being driven by human morality – or the lack thereof. If we could only stop being greedy and selfish. If we could only learn to put other people's needs ahead of our own. Like the Occupy protesters – and I suspect there were not a few environmentalists in that crowd – they know what they want. But they don't know how to get there.

Similarly, I often read material from the social justice movement. Social activists want to stop human trafficking, modern slavery, rape, corruption, child labor. They want a world without poverty. And they agree, we need new values. But they don't know how to get there.

This dilemma is summed up by Gus Speth:

What the authors previously cited and many others are now saying is that today's challenges require a rapid evolution to a new consciousness. That is a profound conclusion. It suggests that today's problems cannot be solved with today's mind ...Today's dominant worldview is simply too biased toward anthropocentrism, materialism, egocentrism, contempocentrism, reductionism, rationalism, and nationalism to sustain the changes needed...

Historian Harvey Nelsen has asked the right question: "How . . . can politics save a culture from itself?" "There is only one way," he answers, "through the development of new consciousness." People have conversion experiences and epiphanies. Can an entire society have a conversion experience? (Speth 2008: 210-211)

These people are looking for God's Kingdom! They don't see it that way, and none of them knows how to achieve it. Nevertheless, they want what God wants: A world that is marked by love, mercy, justice and peace. What they need is to be gently guided toward what we learned earlier: To achieve a world like this we have to start with God. We can answer Gus Speth. A society can have a conversion experience, when the individuals who comprise it become converted. We have to begin with God.

That points us back to our first goal. If we proclaim his name, we will be in position to support God's claim to kingship, and to begin the process of establishing his kingdom.

One thing binds all of those we've been talking about here – the Occupy movement, the environmental and social justice movements, along with their sisters and brothers in similar movements around the world. All of these movements are about human needs. Paul Hawken sees the environmental movement and the social justice movement as two sides of the same human-needs coin, but he thinks he knows which has to come first:

> There is no question that the environmental movement is critical to our survival. Our house is literally burning, and it is only logical that environmentalists expect the social justice movement to get on the environmental bus. But it is the other way around; the only way we are going to put out the fire is to get on the social justice bus and heal our wounds, because in the end, there is only one bus. Armed with that growing realization, we can address all that is harmful externally. What will guide us is a living intelligence that creates miracles every second, carried forth by a movement with no name. (Hawken 190)

In reminding us that 'there is only one bus', Hawken is asking for a single movement of reformation and restoration that will really meet human needs.

That's what God wants, too.

Yes, God wants to proclaim his name.

And he wants to establish his kingdom.

But he also wants to provide for people. And that takes us to Goal Three.

## Goal Three: Provide for His People

'Give us this day our daily bread, and forgive us our debts, as we also have forgiven our debtors. And lead us not into temptation, but deliver us from evil.' Matthew 16:11-12 ESV

'When I look at your heavens, the work of your fingers, the moon and the stars, which you have set in place, what is man that you are mindful of him,and the son of man that you care for him? Yet you have made him a little lower than the heavenly beings and crowned him with glory and honor. You have given him dominion over the works of your hands...' Psalm 8

# 11

## God Cares About Us!

Up to now our discussion has been entirely about God. This is as it should be. Rick Warren started his book *A Purpose Driven Life* with the single sentence: "It's not about you." He was right. Anyone who has had a genuine encounter with God will agree. When you really connect with God, nothing else matters. It's not about us. It's about him.

But when we finally get our eyes off ourselves and focus on God, something surprising happens. The deeper we go into who God is, the clearer it becomes that at the center of God's being is a deep, deep love – for us.

In verse 3 of our psalm, we get an idea of where David might have been when he wrote the poem we call Psalm 8:

*When I look at your heavens, the work of your fingers,*
*the moon and the stars, which you have set in place...*

We moderns live with streetlights and shopping malls. We almost never see the sky the way David saw it every night of his life. You can imagine David out on a hillside, tending sheep when he was young or on a nighttime walk when he was old. The magnificence of the heavens has overwhelmed his imagination. The stars are so bright he can almost use them to count his sheep. The beauty of the sky draws his thoughts to the glory of God, and makes him wonder why a God who can make beauty like this should care about human beings at all:

*what is man that you are mindful of him,*
*and the son of man that you care for him?*

Why should God even notice us, let alone want to do anything for us?

## Nothing but a rounding error

The prophet Isaiah comes to a similar conclusion:

> *Surely the nations are like a drop in a bucket;*
> *they are regarded as dust on the scales;*
> *he weighs the islands as though they were fine dust...*
> *Before him all the nations are as nothing;*
> *they are regarded by him as worthless*
> *and less than nothing.*
>
> Isaiah 40:15,17 NIV

It is easy enough to say that as an individual person I am nothing before God. Who has not felt small and insignificant when confronted with one or another of the wonders of creation? Isaiah's comment is broader than that. The entire human enterprise, everything humanity has ever done or might do, the sum total of our existence all added together is less than a rounding error – dust on the scales – in the accounts of God and his creation.

Like David, Isaiah finds inspiration from meditating on the stars in the sky:

> *Lift up your eyes and look to the heavens:*
> *Who created all these?*
> *He who brings out the starry host one by one*
> *and calls forth each of them by name.*
> *Because of his great power and mighty strength,*
> *not one of them is missing.*
>
> Isaiah 40:26 NIV

And like David, he draws hope rather than despair from the enormous gulf between who God is and who we are:

> *Why do you complain, Jacob?*

*Why do you say, Israel,*
*"My way is hidden from the LORD;*
*my cause is disregarded by my God"?*
*Do you not know?*
*Have you not heard?*
*The LORD is the everlasting God,*
*the Creator of the ends of the earth.*
*He will not grow tired or weary,*
*and his understanding no one can fathom.*
*He gives strength to the weary*
*and increases the power of the weak.*
*Even youths grow tired and weary,*
*and young men stumble and fall;*
*but those who hope in the LORD*
*will renew their strength.*
*They will soar on wings like eagles;*
*they will run and not grow weary,*
*they will walk and not be faint.*

Isaiah 40:27-31, NIV

And here is David's conclusion:

*Yet you have made him a little lower than the heavenly beings*
*and crowned him with glory and honor.*
*You have given him dominion over the works of your hands;*

It is perhaps one of the most unexpected things we discover about the God of the Bible, that his heart beats with a passion for human beings. He really does love us and care about us!

### The order is important

It is important that we remember that this idea (how much God loves us) has only come up halfway through our study. Yes, God does care about us. Yes, God wants his best for us. But God's name and God's kingdom come first. God always provides for us within the framework of his other goals for creation. The order is important.

This is not God being jealous of his rights. This is just how things work in God's creation. It goes back to the fundamental question of relationships. If God's name is not being honored – if we are not living in relationship with God as Creator – our needs cannot be met. If society is not operating according to the values that mark the Kingdom of God, our needs cannot be met.

God's first two goals, to proclaim his name and establish his kingdom, are necessary preliminaries. Yes, human needs are important to God. But human needs can only be satisfied in a world that is proclaiming his name and establishing his kingdom.

### Needs and needs – what kind are we talking about here?

We probably should clarify one more point before we move on. When we use the word "need", most of us probably think of what the dictionary might call "something that is wanted or required": Food. Clothes. Shelter. Medical care. These are all things that Jesus had in mind when he said,

> Therefore do not be anxious, saying, 'What shall we eat?' or 'What shall we drink?' or 'What shall we wear?' For the Gentiles seek after all these things, and your heavenly Father knows that you need them all. (Matthew 6:31-32)

As an aside, notice the very next verse. Jesus concludes this with an admonition that goes to the heart of what we've been talking about here: "But seek first the kingdom of God and his righteousness, and all these things will be added to you." The order is important!

We can take it as established, then, that God knows of our practical needs, and promises to provide them for us. The needs addressed in Psalm 8, however, are on a different level altogether. They have to do with *what we are, who we are, and what our primary task in life is*. These needs aren't practical in the same way that food and clothing are. They are unlikely to be found in even the most effective anti-poverty programs. But having these met – or not – makes all the difference in how satisfied or fulfilled we are as human beings. Being satisfied at the soul-level, for that is what we're talking about, might be as important in the long run as a good meal or a warm bed. Why else would mystics have willingly given up food by fasting and living for weeks or months in the wilderness?

God is happy when we have enough to eat. He wants us to be sheltered from the cold and rain. But his deepest desire is that we become all that he has created us to be.

On this deep level then, what does God want for us? Let's find out.

# 12

## A Unique Position

*Yet you have made him a little lower than the heavenly beings*
*and crowned him with glory and honor.*
*You have given him dominion over the works of your hands;*
*you have put all things under his feet*

We human beings are unique among all of God's creatures. God created us as hybrid creatures. We are both "physical" and "spiritual". We have one foot in two different worlds at the same time. We are animals who can compose symphonies. We're heavenly beings with taste buds!

God, a purely spiritual being, intent on creating a physical universe, wanted to put a creature in this new physical, material world who would also be at home in the realm of the spirit. A creature who would be unlike anything else in the material or spiritual worlds, uniting in himself these two dimensions of existence. And so he made human beings.

### 'Incarnate creatures'

Terminology is a challenge in a discussion like this. I'm satisfied with the label 'material' for our physical aspect; I struggle with what to call our non-material side. "Spiritual" is the best I can offer right now, keeping in mind that what we're trying to label includes everything in us that transcends our physical bodies – self-awareness, consciousness, emotion, rationality, will, up to and including our eternal souls themselves.

And we need to keep in mind that though we may divide ourselves this way in order to explore the topic logically, such division is artificial at best. We are hybrid, but we are hybrid unities. It is not so easy to separate into parts a creature that God has created as a unified whole. A recent book by Matthew Dickerson, *The Mind and the Machine: What it Means to be Human and Why it Matters*, has been helpful for me in understanding what is going on here. Dickerson draws on a number of thinkers including JRR Tolkien as he explores the question of what it means to be both material and spiritual. Here he summarizes a discussion in *The Silmarillion* on the similarities between elves and men (The words in italics are in Tolkien's invented elvish language):

> There are on Earth "incarnate" creatures, Elves and Men: these are made of a union of *hroa* and *fia* (roughly but not exactly equivalent to "body" and "soul"). ...
>
> *Hroa* and *fia*...are wholly distinct in kind, and not on the "same plane of derivation from Eru [God, The Creator]," but were *designed each for the other to abide in perpetual harmony*. The *fia* is indestructible, a unique identity which cannot be disintegrated or absorbed into any other identity. The *hroa*, however, can be destroyed and dissolved: that is a fact of experience. (In such a case he would describe the *fia* as "exiled" or "houseless.") (Dickerson 124. Emphasis mine)

The operative phrase for our discussion is the one I highlighted above. Body and soul are "designed for each other to abide in perpetual harmony." We are not spirit beings who inhabit bodies. Nor are we bodies that have attained self-awareness as an extension of our brain's organic cognitive functions. We are a spiritual/physical unity, 'incarnate creatures' (Tolkien's term) or perhaps better, 'embodied selves'.

### A gift with consequences

This is a gift from God. It sets us apart from every other creature, and is very close to the center of what it means to be a human being. But this gift is a two-edged sword. Living in two worlds gives us amazing experiences and abilities, along with tremendous frustration and tension.

For example, my wife and I share our home with a small white dog named Pumpkin, an elderly Bichon Frise. She's cute and likeable, though not particularly intelligent. Pumpkin looks for three things in life: walks, treats and naps. If she can have two or three walks a day, a treat when she sees the mailman or some peanut butter when a human being has a finger to be licked, and can sleep the rest of the day away, she is perfectly content. Life is good. Whoever coined the phrase, "it's a dog's life" was not describing Pumpkin's experience. I am sometimes jealous; it seems so easy for her to achieve complete fulfillment. She gets 100% every day of her life.

I, meanwhile, start off each day sometimes admiring the sunrise but more often preoccupied with the hours ahead. I have a mental list of the things I need to do, things other people want me to do, things I would like to do some day when I get the time. This list grows longer as the day goes by. Starting the car, I remember that it is overdue for an oil change. As I back out of the driveway my brain registers that the lawn needs to be mowed. Passing the bank, I remember that I haven't paid the mortgage yet. Sitting down to finish a chapter like this one, I find myself instead reading emails that need to be answered. The day comes to an end with more unfinished tasks than finished ones. Pumpkin gets 100%; I'm lucky to end a day at 25%. There are times when I would love to trade places with her!

On the other hand... While it is true that she has no frustration, no list of things to do or guilt over things left undone, she also has no idea what music is. She can't appreciate a sunrise or a sunset. Fashion means nothing to her. Fine cuisine would mean even less. A dead squirrel is more attractive than the best Wisconsin cheese. Progression of time – days, weeks, months, years – is meaningless.

She is as happy and fulfilled as a dog can be. But she is still only a dog.

On the other hand, every human being, even someone with severe physical or mental limitations, experiences life in dimensions that a dog can't begin to fathom. This is because we have what a dog does not have, our hybrid "embodied self" nature. We experience life through our physical senses, just as a dog does, but we process those experiences with minds that imagine, anticipate, project, feel and create. An ability to taste, imagine and manipulate material things becomes an exquisite Italian meal. As we hear, feel, and create, sounds become a symphony. Movement around a room with another person becomes a dance. Two sets of eyes gazing at each other, two hands gently touching become a life-long love affair.

God made us this way – a little lower than the angels, a little higher than the beasts. And if he made us like this, it follows that he wants us to understand and develop this unique dimension of our existence. He wants us to experience all the joys of physical existence – the beauty of a sunset, the shock of diving into a cold pond on a hot summer day, the satisfaction of rest after a long day of work in the sun. This means we need to use our bodies – and to keep those bodies healthy with good food, exercise and rest.

But he also wants us to experience the thrill of using our brains to solve difficult problems, the satisfaction of producing a painting, the pain and joy of human relationships. And he longs for us to have the supreme experience of doing all of this in the context of a growing relationship with God himself through the influence of his Holy Spirit in our lives.

### Some assembly required

Our hybrid nature is a gift from God. But it is a gift that comes to us undeveloped, like a toy in a box with the label, 'Some assembly required'. This dimension of our existence has to be recognized and cultivated if we are to reach our full potential. Living as fully developed incarnate creatures does not come automatically.

But where does this happen?

There is solid research now that suggests that it is in nature, that is, in God's creation, that all of this comes together. Nature is the setting in which we best experience the physical, intellectual, emotional and spiritual aspects of our existence. Whether it is the quiet of a forest glen, the roar of an ocean beach or the peace of a backyard garden, these are the places where we experience what it really means to be a human being.

Richard Louv's *Last Child in the Woods* (2006, 2008) and *The Nature Principle* (2011) are a good place to start exploring this topic.

A growing body of research links our mental, physical and spiritual health directly to our association with nature – in positive ways. Several of these studies suggest that thoughtful exposure of youngsters to nature can even be a powerful form of therapy for attention-deficit disorders and other maladies. As one scientist puts it, we can now assume that just as children need good nutrition and adequate sleep, they may very well need contact with nature. [Louv 2008: 3]

I have recently noticed more pictures of nature, and even televisions showing nature scenes, in the halls of medical buildings. People have been reading this research! Unfortunately, it appears that just looking at pictures of nature isn't enough. We have to actually be immersed in God's creation to receive the full benefit. Research at the University of Washington examined the effect of 'nature' on people by exposing subjects to 'low level stress' and then putting them in front of either a blank wall, a window looking out at a nature scene, or a high-definition plasma video screen showing the same nature scene as the window. Subjects looking out of the window experienced faster recovery (lowered heart rates, for example) than those who had nothing but a blank wall to look at. However, those subjects who were looking at the video screen showed no difference from those looking at a blank wall. God's creation doesn't seem to make it through electronic pixels as effectively as we might think it does. This study might be relevant for churches projecting more and more of their sermons to remote sites on video, or blocking windows that have a view of God's creation in order to show images of that same creation on a screen!

*Last Child in the Woods* is full of examples from schools, hospitals and even prisons that show the benefits people receive, physiologically and psychologically, when they are exposed to nature.

The "where" is important in terms of developing our hybrid nature, but so is the "when". The title of Louv's book – *Last Child in the Woods* – reminds us that development happens as we are growing up. What this points to is a world in which children – young human beings – are able to grow up with a full experience of life, physically, spiritually, intellectually and emotionally.

God wants all of this for us, and his plan was that we would run his world so that this fundamental need can be met. Let's put this into another of our management principles:

**Principle #4: In a world that reflects God's goals, every human being will be able to live a life that is physically healthy and intellectually, emotionally and spiritually rich and fulfilling.**

## Money isn't everything

Think about how far we are from a world like that right now. A large proportion of our work force – office workers, for example – get very little physical exercise and work at jobs where our mental processes are stretched beyond capacity. We work in a cubicle all day, and carry work home on a smartphone at night. Physical exercise is an effort, and sometimes a scheduling nightmare. Our lives are lived under artificial lighting in controlled climates where we are surrounded by high-tech equipment and electronic gadgets. Our encounters with nature are limited to a vase of flowers, window views of landscaped lawns or perhaps an insect that managed to invade our technological cocoon.

We maintain a blissful ignorance of the fact that our artificial existence depends on the labor of millions of other people doing mindless physical jobs – migrant field workers, for example, who pick and pack the vegetables we buy on the way home from our cubicles. These people work long hours for little pay in hot fields. They are getting more than they want of the great outdoors, but their opportunity for intellectual or spiritual stimulation is as limited as the office worker's chances of getting decent physical exercise.

I'm not going where you might think I am with this argument. My concern here is not the economic disparity, though if the world were operating on kingdom values, there would be far less than we see today. Our section on God's Kingdom earlier speaks implicitly to this issue as well as many others. No, the amount a person is paid is not what is at issue here. Nor is it a perceived unfairness that some people do hard physical labor while others sit in a chair all day. There is nothing wrong with physical labor. Human beings have been doing hard work for 10,000 years and many doctors will tell you that the person using his body actively is probably going to be healthier than the cubicle dweller.

No, the problem is that there is a truncated experience of life at both ends of the spectrum: Both the knowledge worker who is not able to use her body and the laborer who cannot develop his mind are being deprived of the life God wants them to have.

We need, and God wants us to have, a world in which all of us, whatever our economic status, can have lives lived fully in all dimensions – a world in which those who work primarily with their hands and bodies are paid fairly, but also have opportunity to develop their minds and souls. And one in which those who work primarily with their heads do so in an environment that is physically stimulating and healthy as well. Such a world is not impossible to imagine.

But God has even more in mind for us than this.

# 13

## A Special Status

*Yet you have made him a little lower than the heavenly beings
and crowned him with glory and honor.
You have given him dominion over the works of your hands;
you have put all things under his feet*

Our hybrid nature speaks to what we are. The next gift from God addresses who we are. These two aspects of our lives are closely connected. Our hybrid existence is not an arbitrary design decision. It is both the source and result of what David calls glory and honor with which we have been crowned.

Our glory and honor is in this:

Then God said, "Let us make man in our image, after our likeness. And let them have dominion... So God created man in his own image, in the image of God he created him; male and female he created them. (Genesis 1:26, 27)

We are made in God's image. But what does this mean? Of what does this image consist?

### We reflect God's nature and being

Part of the image has to do with our being: who we are is connected to what we are. Thus the image is in part an indication that what we are reflects what God is.

We are arguably the most advanced creatures in creation. Our ability to reason, to communicate both concrete and abstract thought by use of language, to create works of art and tools and technology of all kinds, all these and more are signs of God's image in us. These are the 'glory and honor' with which we have been crowned.

We should be careful not to define this crown too narrowly. For example, tool-making used to be considered a uniquely human ability, until Jane Goodall's ground breaking research identified tool use among chimpanzees. Now we've even seen evidence of some birds using tools in a very rudimentary fashion. Language? Apes have been taught to use sign language to communicate. I just saw a report of primates using iPads! We know of sophisticated communication among whales and other sea mammals that may come close to what we call language. We have faculties that are clearly not superior to those of other creatures. A bird's eyesight and a dog's sense of smell are both far better than ours.

This crown of the image of God is not a single ability or even a group of them. What we have from God is a package that we can call human nature, a bundle that ties the material and nonmaterial aspects of our being together into a unity. We have the embodied self of which Matthew Dickerson spoke that in some mysterious way reflects God's own nature. We "look like" God, not physically, of course, but in the way in which our bundle of traits and abilities matches his own. We communicate because God does. We create because God is a creator God. We love beauty, harmony, order because God loves these things. We have an ability and a need to love others and to be loved by others because God is a relational God. We are reflections. Our crown of splendor and honor is found our likeness to God himself.

But that's not all.

## We represent God's interests

There was a custom in the ancient world that the author of Genesis may have had in mind when he wrote his account of our creation. Kings who had conquered a city would sometimes set up statues of themselves in the central square as a symbol of their authority. In a vast empire with slow travel and no modern media, most of a king's subjects would never actually see him in person. But the statue or image would be a daily reminder to the citizens of that town that they had a king who ruled over them.

In the same way, Adam and Eve were created to live in creation as God's representatives. And if true of them, it is true of us today. Made in God's image we are image-bearers: If our human nature – our abilities and our relational capacities – in some way reflects God's nature, so our very presence in creation represents God's authority.

Remember the ranch analogy we used back at the beginning? That is what we're talking about here. Being God's managers is another way of describing this business of representing God's interests.

**Putting on the crown**

If God has given us this crown of glory and honor, we have to wear it. This means we have to understand what being an image-bearer means, and more importantly, live as if we were the royal people that we are.

Perhaps you saw the film "The Help". This film is about a group of black maids in the southern United States in the 1960's. It is a powerful social commentary on a difficult period in US history, but it also has an important, if indirect, message about parenting. In a household where the children were emotionally neglected by their parents, maid Aibilene takes it on herself to raise "her babies" with a strong message of self worth. "You is Kind, You is Smart, You is Important" she repeats over and over, hoping that the child will hang on to this vital message of self-worth. When I was growing up, I recall a relative who may have struggled with similar issues, for she had a paper on her bathroom mirror: "God don't make no junk."

To wear the crown of an image-bearer we have to think of ourselves as such. We're not just children of God – we're co-rulers with God in creation and beyond.

The Spirit himself bears witness with our spirit that we are children of God, and if children, then heirs—heirs of God and fellow heirs with Christ. (Romans 8:16-17 ESV)

This leads to our next principle:

**Principle #5: In a world that reflects God's goals, every human being will understand that he or she is created in God's image, and will be treated as such a person deserves to be treated.**

I think you will agree that we're a long way from a world like that right now.

This principle depends on the earlier ones. "Every human being will understand that he or she is created in God's image" assumes that all of these people have had the opportunity to hear the gospel – God's name has been proclaimed to them (Principle #1). And to "be treated as such a person deserves to be treated" assumes a society that is operating according to the kingdom values of love, justice and mercy (Principle #3).

The implications of this principle are profound. For the world to work this way, every social system that is built on discrimination of any kind would be upended. People would still do "menial" jobs – the trash has to be taken out and the toilet has to be cleaned no matter how ideal our world is. But those who do such tasks would receive as much respect and honor as those who do less menial (and often less necessary) jobs. No longer would a person take home millions of dollars because he can put a ball in a hole while those doing more important tasks have trouble meeting basic needs. An athlete's place in society would sink dramatically compared to the work of those who nurse the sick, or care for children.

This is starting to look like a major challenge, isn't it? How might such a society come about? Who is going to build it? We are!

# 14

## An Awesome Responsibility

*Yet you have made him a little lower than the heavenly beings*
*and crowned him with glory and honor.*
*You have given him dominion over the works of your hands;*
*you have put all things under his feet*

We've seen how God cares for us in what we are – hybrid material/spiritual beings. We've seen how he has blessed us with glory and honor in who we are – bearers of his image. Now we see the task he has given us – responsibility over his creation.

This is really where we started our discussion. Think back to our ranch analogy, where we first came across this line in our psalm. This line is in fact our Power of Attorney document – it proves that God has given us the keys to his creation, that we have authority from him, that we can sign the checks.

There are a couple of things we need to note about this.

### Responsibility before privilege

The subject is dominion, authority – something most of us think of as a privilege. But I'm calling it responsibility, which sounds like more of a burden. There is a reason for this. In giving us dominion over creation, God was not rewarding us for something. No! He was giving us a job to do. We have a task, like the management of a ranch, from God and for creation. It isn't about us –

it is about the One who gave us the job and about those whom we serve as we go about this task.

In high school I was a bagger at a supermarket. I stood at the end of the conveyor belt and put groceries in bags for customers. I often took the bags out to their cars as well. In the small world of that supermarket, the baggers (we were all teenage boys) were at the bottom of the social scale. Above us were the checkers – almost all girls. I suppose girls were just better at remembering prices and pushing buttons (no scanners, of course). The checkers, in turn, were below the stock boys who spent their time restocking the shelves, and then department managers. At the very top, ruling over this kingdom from an office high above the line of cash registers, was the Store Manager.

To the teenage workforce, the Store Manager was like God. We could hardly imagine the privileges that must come with a job like his, and we were sure he was wealthy beyond dream. As far as our lives in the store, he had the power of life and death over us. He determined work schedules and assignments – who got the fun jobs and who had to clean the restrooms. He could promote us. He could hand out raises and bonuses. He could fire us. A visit to his office was about as traumatic an event as we could imagine.

There was another side to his job, though – one that I know now, though I was oblivious to it back then. He had responsibilities that we teenagers knew nothing about. He had to be sure the store was opened on time; he was almost always the one who locked up at night. If someone didn't show for a shift he had to figure out how to adjust staffing. If there was a snowstorm, he had to decide if we would stay open or close up. If merchandise was stolen, if sales were down, it was all on his shoulders.

Knowing what I know today, I would have to think long and hard before I accepted a position like his. There seems to be a lot more responsibility than privilege attached to the title of Store Manager.

As a bagger I looked at the manager's office and saw benefits. He looked out at the store and felt the burden of responsibility. And this is how we have to see God's "gift" of dominion over his creation. A privilege, yes. But also, and much more, a responsibility.

## Responsibility with authority

One of the things that made the manager such an important figure in the world of the supermarket was the fact that he had real power. He could hire, fire, and set hours and wages – his decisions had real consequences for us: more or less money in our pockets. If we slacked off on the job and he found out about it, things weren't going to go so well. What we didn't understand was that he had authority because he had responsibilities. If he was to do his job, he had to have authority.

In the same way, our responsibility for creation comes with authority. We have real power over creation: We can kill animals. We can dam rivers. We are able to blow mountains apart. But, like the store manager, we need to remember that this authority has a purpose. We have it so that we can do the task of ruling over God's creation.

Maybe I can explain better by changing the metaphor.

Not long ago I was presenting some material on creation care to a class of junior and senior high school students. The class was a double-length ninety-minute session that began at 8 am, when most teenagers are barely functional. To make things worse, the room was warm, and it wasn't too long before heads were nodding. I had the sinking feeling that comes to a speaker when he senses that his audience is drifting away.

What to do? I had already tried a joke or two, and those had gone flat. Then I remembered that my next slide had a number of important pieces of information listed, and as the list came up on the screen, I said, "You might want to write these down – they could be on the test..." Instant response! Eyes were open, notebooks were out, pens were moving.

What had I done? I had exercised authority as a teacher. Even though I wasn't going to be in that class long enough to give them a test, their regular teacher was in the room, and they knew it was very possible that this material would indeed show up on a midterm or a final exam.

Let's think about this for a minute. A teacher has authority – real power – over his students. It isn't monetary, like the store manager – but he can still deliver consequences that matter. A trip to the principal's office. A note home to Mom and Dad. A failing grade that could affect what colleges might be available. Why does the teacher have that authority? It is not an arbitrary gift from the administration, is it? No, he has it so he can do his job, which is to deliver education – knowledge and skills – to his students. Without authority, he can't be an effective teacher. But he must always remember the reason he has this power. That reason should always guide how he uses it and what he uses it for.

In the same way, God has given us authority over his creation. And he has given it to us for the same reason: So we can do the job of caring for it. We would be less likely to abuse our power over creation if we were to remember the reason we have that power.


### "All things"?

Our authority is real. And it is comprehensive. Look again at how our dominion is described in the psalm:

> *You have given him dominion over the works of your hands;*
> *you have put all things under his feet,*
> *all sheep and oxen,*
> *and also the beasts of the field,*
> *the birds of the heavens, and the fish of the sea,*
> *whatever passes along the paths of the seas.*

The geographical progression is important and instructive. God has put 'all things' under our feet: Sheep and oxen represent domestic livestock – the animals right outside our door. The beasts of the field suggest wild animals that can be hunted. The birds of the heaven and the fish of the sea are one step further removed. You could hunt birds and catch fish, but their worlds were mysterious and unknown. And whatever passes along the paths of the sea seems to be a phrase that captures everything else in creation. The sea was a vast unknown – even today we know more about the surface of the moon than we know about the depths of the ocean. But our God-given dominion is said to extend even to this unknown and mysterious realm.

Note, though – God is not talking about human control of every creature. Dominion does not mean that every creature can or should be domesticated. If there were any doubt about that, a quick look at the book of Job might be helpful:

> *Is the wild ox willing to serve you?*
> *Will he spend the night at your manger?*
> *Can you bind him in the furrow with ropes,*
> *or will he harrow the valleys after you?*
> *Will you depend on him because his strength is great,*
> *and will you leave to him your labor?*
> *Do you have faith in him that he will return your grain*
> *and gather it to your threshing floor?*
>
> *Can you draw out Leviathan with a fishhook*
> *or press down his tongue with a cord?*
> *Can you put a rope in his nose*
> *or pierce his jaw with a hook?*
> *Will he make many pleas to you?*
> *Will he speak to you soft words?*
> *Will he make a covenant with you*
> *to take him for your servant forever?*
> *Will you play with him as with a bird,*
> *or will you put him on a leash for your girls?*

Job 39:9-12, 41:1-6

God never intended these creatures to be controlled by us, but they are still included in the circle of our authority. We are responsible for them. Their welfare is our concern.

**Responsibility with consequences**

There was one aspect of the store manager's job that I knew nothing about. I was oblivious of his job performance criteria. I knew he could punish or reward me for my performance; it never occurred to me that he could be rewarded or dismissed for his. His dominion over that store came with a price. He had to perform.

So with us. We don't get much light on the performance criteria that God has for us from Psalm 8, but Jesus later told a parable that applies perhaps too well. It is commonly called the Parable of the Talents, and is found with minor variations in both Matthew 25 and Luke 19. A king goes on a journey, leaving his servants with varying amounts of money – ten talents, five talents, one talent. We have no way of knowing how much money was involved, but even one talent was a considerable sum, perhaps as much as $100,000 or more in today's currency. While the king is gone, the first and second servants invest the money, and double it. The one who started with ten now has twenty and so on. But what happened to the servant with one talent? Being highly risk averse, he buried the money in his backyard! All he had to bring back to the King on his return was the original amount.

The story ends with rewards handed out to the faithful servants, and harsh punishment for the last one. This parable has been used, appropriately, to remind us that all the "talents" – gifts and abilities – that we have are from God and that he expects a return on his investment in us. It is a sin to 'bury your talent', whatever that might be.

So far, so good. But what if the parable applies corporately as well as individually? What if we consider the wealth represented by creation itself as something that the king has left in our care? When the king in the parable returned he was angry at his servant, even though he had lost no money. He was expecting a profit! When he got only his principal back, the King was upset, and that was the end of that servant.

Let's go back to Paul Hawken and his speech at the University of Portland. Do you remember how he began?

Class of 2009: You are going to have to figure out what it means to be a human being on earth at a time when every living system is declining, and the rate of decline is accelerating.

Hawken is right. By almost any measure, God's creation is in serious decline. In the context of the parable, the king is not only not going to see a profit, he's going to see a major decline in the value of his property. We're much worse off than the servant who buried the talent; no, we have been living like the Prodigal Son (see Luke 15:11ff), squandering our master's money on "reckless living" without realizing that he wants it all back – with profit.

So what do we do?

## Doing dominion right

These lines we have been looking at about dominion are the climax of the psalm. Everything we've been studying has been leading up to this: God has prepared us to rule his creation. So, the last of our principles:

**Principle #6: In a world that reflects God's goals, every person will understand that he or she shares responsibility for ruling God's world, and will have the opportunity to exercise that responsibility in his or her own life.**

We abuse God's creation partly because we've forgotten the reason for the authority we have. The servants needed to understand that the King's money was in their hand so that they could earn a profit for their master. A good teacher is one who remembers that the authority she has in the classroom is always and only to be used to accomplish the goal of helping students to learn.

It is easy to focus on the foolish servant who buried the money, or to remember bad teachers who abused their authority as mini-dictators in their classrooms, or who failed to exercise authority at all, and so accomplished no real education. But what about the others? What could be more satisfying than to be one of those successful servants, hearing the King's praise, and being told, "Because you have been faithful in a very little, you shall have authority over ten cities."

**A virtuous cycle**
We've seen in both the Lord's Prayer and Psalm 8 that God cares about

## Proclaiming His Name

which results in God

## Establishing His Kingdom

which ultimately makes possible

## Caring for People

When we zoom in on the last item in the list using Psalm 8, we find that God's care for people appears in

What we are    Incarnate beings

Who we are    God's image bearers

What we do    Care for God's creation

Psalm 8 starts and ends:

O LORD, our Lord… O Yahweh, our Sovereign Master…

Our dominion over God's creation runs on the rails of God's dominion over us. We rule as he rules. Our goals have to be his goals. This means that we will accomplish our task of ruling over God's creation as we proclaim his name, establish his kingdom, and care for people.

And we'll care for people by having a world and a society where human beings understand their unique position (incarnate beings), live up to their special status of being God's image bearers, and participate in the task of caring for God's creation.

And people will participate in caring for God's creation by proclaiming his name, establishing his kingdom, caring for people...

This is a virtuous cycle. A vicious cycle is one in which each element reinforces the others in a negative direction. Here every element reinforces the others in a positive direction.

As God's name is proclaimed

...*verbally*, people are able to have a relationship with God and to develop that relationship with other people around them;

...*biospherically*, through a flourishing biosphere, people will be healthier, happier, better able to proclaim his name verbally – in witness and worship.

As God's kingdom is established with love, mercy and justice, society becomes safer, more peaceful and more equal.

As people are cared for, it is more likely that they will live in families and communities where they not only have their physical needs met, but also know

...*what they are* (material/spiritual incarnate beings),

...*who they are* (images of God) and

...*what their task in this world is* (to proclaim God's name, establish his kingdom, care for people – which takes us back to the beginning as the cycle repeats itself).

Is it possible to have a world that looks like this? Let's find out.

# Part III
## Getting There from Here

IN WHICH WE BEGIN TO CONSIDER HOW TO TAKE THE
TRUTHS WE'VE LEARNED AND PUT THEM TO WORK IN
THE REAL WORLD.

# 15

## A Place to Begin

Where do we begin?

We've developed a nice formula: Name, kingdom, people. We've come up with a series of objectives that, if implemented, would transform our world. But all we have to do is crack open the door or peek out the window at the world we live in – and it all seems hopeless. Where do we even start?

**Let's do a puzzle!**

I am part of a large extended family that loves games and puzzles. At Christmas gatherings or family reunions, there are always board games in progress – and there is often a large jigsaw puzzle being put together as well. These are serious puzzles of 5,000 pieces or more, and the family members who like them are serious puzzlers. A group of four or five people will be huddled over a card table, sorting pieces into similar color patterns, sifting, scanning, trying to fit one piece into another, rejecting it and trying again with another.

At the beginning the task looks impossible. Other members of the family look on with bemusement and then return to their own chosen pastimes – arguing politics, watching football, reading mystery novels. But the puzzlers keep at it, piece by piece, and gradually the puzzle begins to take shape. The border is complete. A lake appears in the lower left, puffy clouds form in the sky portion. As the pieces find their places, the pace quickens and eventually there are just a handful of gaps in the picture – and finally one person ceremoniously places the last piece. The puzzle is finished.

The puzzlers congratulate each other, admire their work, grab some more hot chocolate – and likely as not, dump the puzzle pieces back in the box and start another one. Such is the nature of their passion, or as some would say, their addiction.

Addicted or not, these folks offer some lessons for the rest of us as we think about putting our world back together.

Puzzlers tend to specialize: One person sorts through the blue pieces, another finds everything that looks like a tree, a third concentrates on pieces that seem like they are part of the barn in the middle. But they also have to cooperate. That blue "sky" piece might well turn out to be part of the lake. The obvious "tree" piece has, on closer examination, part of the barn wall on it as well. I find them constantly poking around in the pieces collected by another person, just in case one of their sky pieces has been mixed in with those of the lake. And they have to be persistent and patient. Puzzling is not for those who need instant gratification – in fact, the bigger the challenge and the longer it takes the more satisfied these people are.

A lot of us are working on the world's problems in many different areas. There are so many books out now in ecology, economics, political science, sociology and a host of other subjects that could relate to what we've been studying that it is impossible to keep up. The people who go deep in these areas are like the puzzlers who decide to concentrate on one color or one section of the picture. They know a lot about their own areas; many are genuine experts. But the more they know about their own sections, the less they are likely to really understand other parts of the problem. We all need each other.

That is why, like good puzzle solvers, we need to learn how to cooperate and learn from each other. We need to learn to look over each other's shoulders, as it were, and learn to play with other parts of the puzzle just to see what might fit.

### It's going to take a while

With all of this, we're going to have to be patient with each other and with the size of the task in front of us. It's going to take time, as I learned a few years ago while attending a conference in Jamaica. Late one afternoon one of the conference attendees, a local professor, invited two of us to visit an unusual garden a short distance away from the hotel where we were staying. We drove about 15 minutes, parked and began to walk into a forest and then along a flowing mountain stream. I looked around, and realized that I was in the middle of a genuine tropical forest that appeared to be hundreds of years old. It felt like we were hundreds of miles from the tourist developments right outside the garden gates. Here were enormous trees in every direction; the stream ran flowing over moss-covered rocks. And there were birds – such birds. It was a glorious experience.

You know what made it even more amazing? It turned out we were walking on a ruined coffee plantation. Twenty-five years before, a Christian couple had purchased an economically worthless piece of ground for almost nothing. The soil was so bad that the real estate broker had tried to talk them out of buying the place. But they had a dream of taking a ruined piece of God's creation and turning it into something beautiful again. They succeeded – how they succeeded! It took them a quarter of a century, but when you walk through the property now, you could easily think it had grown up untouched by humans for centuries.

I took a couple of things away from this experience. I was encouraged that such a miraculous transformation from ruin to beauty is even possible. It seems that God has placed within his creation an astounding capacity for regeneration. We ought not to allow ourselves to get to discouraged – in the right hands, miracles can happen. And that was the second lesson: This ability creation has to recover from abuse seems to be triggered when someone comes along who really loves it and who knows what he or she is doing. But I was also sobered by the time commitment required. I have had informal conversations with other people in other places who are engaged in the same kind of rescue operations. They all suggest the same thing: Twenty-five years seems to be an average. God's creation can bounce back, but it will take a different time horizon than you and I are used to.

What, then, am I asking you to do? You probably are not in a position to buy a ruined coffee plantation – though you might have a piece of family property that you can take under your care with the idea of seeing what you can do to heal and restore it (or to keep it beautiful) with God's help, to see what God can do with it with your help. There are many other ways, though, that you can apply the lessons of this book in your own world.

Here are some thoughts to get you started:

## Zones of influence

Think about those parts of your life where you can begin to make an impact. Let's call them "zones of influence." I am thinking about areas like home and family, work, community and social life, and church – places or groups of people that you interact with on a regular basis and where people will listen to what you say. Pick two or three areas where you might have the greatest opportunity to influence others or to shape decisions that are made for the group.

This list will be different for each of us, and will depend on all kinds of factors. Your stage in life matters. Young adults without family obligations have time and energy, but may not be in a position of significant influence at work. A couple in upper middle age (where I am!) has considerably less energy, but is much more likely to be in a position of influence in their work setting or in their church. Don't worry about the number of people you are involved with in this zone. A stay-at-home Mom with two kids is the Number One influencer for those children for at least ten years, and that is not a small thing.

Take a piece of paper and list your 'zones of influence' down the left hand side (see the example). If you have a position of influence in your work setting, you might want to list that first. If you work part-time, or otherwise spend most of your time with your kids at home, home should come first. Some of these might need to be combined, such as if your job happens to be in your church.

## Apply our objectives

The six objectives we developed through this study can be summarized with short labels:

Verbal proclamation – proclaiming God's name in prayer, worship, out-reach.

Biospherical proclamation – causing the biosphere to flourish.

Kingdom values – promoting love, justice, and mercy.

A balanced life – developing all dimensions of life (physical, spiritual, intellectual).

Respect for all – treating human beings as if they were images of God (because they are).

Dominion – participating in the rule of God's creation by living out these objectives and helping others to do so (the virtuous cycle begins...).

Now add the first five objectives to your chart, as in the following table. We are not including the last objective, since it circles back and encompasses all of the others.

Table 2

| | Management Objectives | | | | |
|---|---|---|---|---|---|
| Zones of Influence | 1. Verbal Proclamation | 2. Biospherical Proclamation | 3. Kingdom Values | 4. Balanced LIfe | 5. Respect for all |
| Home | Action items .. | | | | |
| Work | | | | | |
| Church | | | | | |
| Social Life | | | | | |
| etc. | | | | | |
| | | | | | |

Now list things that you can do in each zone of influence to advance each of these objectives. You will discover quickly that in some zones of influence, some of the objectives don't apply easily. For example, you might have a work situation such as teaching in a public school where you cannot speak publicly about your faith. That's okay – be creative!

## Fill it in with action steps

Now we can get specific. For each row ("zone of influence") go across and write in two or three action steps you can do in this area to advance that particular objective.

Let's make a sample chart for a recent college graduate living with roommates and working in a starting level position for a large corporation. Her chart might look like this:

Table 3

| Zones of Influence | Management Objectives | | | | |
| --- | --- | --- | --- | --- | --- |
| | 1. Verbal Proclamation | 2. Biospherical Proclamation | 3. Kingdom Values | 4. Balanced Life | 5. Respect for all |
| Home | Maintain an active witness to neighbors. See if roomrates are interested in a house prayer time every week or so. | Be conscientious with recycling, energy use, food, cleaning products, etc. Live as "green" as possible. | Exercise kingdom values in my personal life. Be aware of my tendency to be critical rather than showing mercy to others (esp. Joe!) | Take care of myself physically (exercise!); limit my time online; go to a concert; go to see grandparents | Review my attitudes and actions toward my roommates & others. |
| Work | Be a consistent "quiet witness"; participate in work prayer session on Thursday mornings . . . | Turn off my computer when I leave the office! Volunteer to be on the corporate Energy Review task force | | Try to maintain boundaries betwen work and home life - i.e. get my work done in the office more often! | Be careful how I treat other people at work. Remember that they are all created in His image! |
| Church | Participate in Sunday worship a bit more often, start attending a home fellowship group; teach Sunday School this summer! | Ask coffee-hour people to get us real mugs (or bring my own travel mug!) | Figure out how to emphasize Kingdom values in Sunday School sessions. | | |
| Social Life | | | | | |

Not all of the squares need to be filled in. Some of our objectives don't apply easily in some zones of influence, and that's okay.

So far, so good. If everyone one of us who has read this book were completely faithful in doing the kinds of action items we have listed, the world would be a different place. And I hope you will find it helpful, and that your little corner of God's world will become a nicer place because of your efforts.

But we need to be realistic. This is approaching the problem from the individual point of view, but many of the problems in our world come not from individuals but from the actions of governments and huge multinational corporations.

What do we do about that?

## How do you change a society?

My wife Susanna is my first reader and best critic, and she has reminded me throughout the writing process that oh-so-neat formulas have to make their way into real life. It is fine to talk about changing the world so that it reflects God's goals. The problem is that changing things one-by-one-by-one is an exercise doomed to failure. Unless you are one of the very few people living "off the grid" of modern life, you – like me – are trapped in a web woven by corporations and governments who care little or not at all for the kinds of principles and values we have been exploring throughout this book.

I was thinking about this a few weeks ago on the weekend when we switched from Standard Time to Daylight Savings Time. This nationwide exercise requires that all of us, on one agreed upon night, shift our clocks ahead by one hour, and then reverse in the fall. This is supposed to save energy by moving an hour of daylight from the early morning, when most of us are asleep, to the evening when we're still awake. Even though this has never been proven, we go through this exercise twice a year. 'Spring ahead' in March, 'Fall back' in October.

What if you or I refused to go along? Saying to ourselves, "Standard time is good enough for me," we would keep clocks the way they were. What would happen? Within the walls of our house, nothing would change. NPR News would begin at 2pm instead of 3pm, television prime time hours would start an hour earlier, but these would just be minor mental adjustments that we could get used to. Our real challenges would start as soon as we left the house. We would have to be at the office at 8am rather than 9am, though we could leave at 4pm instead of 5pm, according to our personal time zone. Setting meetings with other people would be a challenge. To get to a plane on time, we would have to be sure to tell the taxi to pick us up on his time, not ours. No matter how much we might insist that we were living according to our own time, we would be making so many adjustments to the outside world that we might as well just give in and change our own clocks to match the rest of the world.

We can have any personal time zone we like, but we still have to live by the community time zone whenever we interact with anyone else. And this is exactly the same dilemma we will face if we try to take the lessons of this book seriously. How can we possibly implement the ideas we've been studying when we are embedded in a global economy and a worldwide society that is committed to just the opposite?

**Some of us can make a difference!**

Most of us reading this would consider ourselves "little people". We have limited zones of influence, and there is no way we could begin to affect a national environmental policy or a corporation's ethical practices. Some of us are not little people, however. We are on the boards of major corporations. We are political leaders. We hold positions in media. We're major investors. We do have enough power to affect policies. We could change the behavior of the corporations we work with, if we wanted to do so.

There are some already doing this. I've even met a few. Several years ago Walmart embarked on a sustainability effort that calls for, among other things, 100% renewable energy and zero waste from all of their operations. Where did this change come from? There were people inside Walmart who made it happen. Not long ago I had an airplane conversation with an engineer from Dean Foods, the largest food processor in the United States, in which I learned that his company is making major efforts to reduce the amount of water used in their factories. Dean Foods as an entity may not care about water, but somewhere inside the labyrinth that is that corporation, there are people who do care. And they are making their voices heard.

There are many reasons for such corporate decisions. Some are purely business decisions, and have nothing to do with values or ethics or God. Someone inside Dean Foods has realized that if their factories run out of water, they will be in big trouble. No water, no orange juice, no profits. It's the same with Walmart. People there have figured out that renewable energy sources are less subject to the vagaries of the markets. More stable prices, more predictable costs, better profits.

What interests me is not the motives, but the fact that in each of these situations someone – a person! – spoke up in a meeting or wrote a memo and so changed the behavior of an entire corporation. This is what I mean by using your zone of influence. If those of us who have this kind of influence were to speak up like this, we could begin to change the world.

Will we? Consider Esther. Esther was a Jewish girl chosen to be queen of the Persian empire. When her people were threatened by an empire wide massacre, Esther's uncle sent her this message: "Who knows whether you have not come to the kingdom for such a time as this?" (Esther 4:15)

Perhaps we need to realize that we do, indeed, live in "such a time as this."

# 16

## Preparing for a New World

It is hard not to agree, whatever our historical and political perspectives might be, that our world is in trouble. Economic institutions crumble, political systems deadlock, and environmental problems stack up like an ice jam on a Wisconsin river. World systems are anything but stable. Change is coming.

The only question is when that change will happen and what kind of change it will be.

I am not trying to make a political statement. This is just an observation based on events you and I are witnessing every day of the week. Should we not be drawing conclusions from these events? Jesus told us to:

> When it is evening, you say, 'It will be fair weather, for the sky is red.' And in the morning, 'It will be stormy today, for the sky is red and threatening.' You know how to interpret the appearance of the sky, but you cannot interpret the signs of the times. (Matthew 16:2-3)

Great events were unfolding right in front of these people, but they were unaware. We need to be careful that we don't fall into the same trap. Remember Paul Hawken, Gus Speth and other thinkers we discussed earlier back in Chapter 1? They are among those who are reading "the signs of the times", and they have all come to the conclusion that our present world system is on its last legs.

I began *Our Father's World* citing Jared Diamond's work, *Collapse: How Societies Choose to Fail or Succeed* and it somehow seems appropriate to go back to him as we end this study. After surveying a number of ancient and modern societies from Easter Island to Haiti that have all gone through some measure of disintegration, Diamond summarizes twelve problems that he believes are hanging over our modern, global world. The problems he lists include, among others, the destruction of natural habitat, loss of arable farmland to erosion and abuse, looming energy and water shortages, the accumulation of toxins (pollution) in air and water, and the increase of greenhouse gases in the atmosphere (climate change). You could argue with his specific list if you wanted to. The details, however, are not as important for us as is his conclusion:

> The world's environmental problems will get resolved, in one way or another, within the lifetimes of the children and young adults alive today. The only question is whether they will become resolved in pleasant ways of our own choice, or in unpleasant ways not of our choice, such as warfare, genocide, starvation, disease epidemics, and collapses of societies... (Diamond 498)

If Jared Diamond and these other thinkers are right, then change is coming, and the world as we know it with its economic and political systems cannot last.

**What then?**

As it happens, a scenario such as Diamond sketches for us is completely consistent with traditional Christian teachings about the end times. We know that Jesus will return to bring an end to history. We understand from his teachings that this period of history will be marked by trouble and turmoil. What we do not know, and cannot know is where the boundary line falls. Will Jesus return and rescue us before a major collapse? Will he wait until it's all over and the dust has started to settle? Might he wait even longer than that?

> Therefore, stay awake, for you do not know on what day your Lord is coming. But know this, that if the master of the house had known in what part of the night the thief was coming, he would have stayed awake and would not have let his house be broken into. Therefore you also must be ready, for the Son of Man is coming at an hour you do not expect. (Matthew 24:42-44)

These verses are a reminder that we need to be ready for Jesus to come at any time. True enough. But they also mean that we need to be prepared for him not to come at any time. Since we do not and cannot know which it will be, our strategy for the future necessarily requires that we consider how to live through changes – even collapses – that might overtake us.

Whether Jesus comes back soon, or leaves us to muddle through for a few decades or centuries more doesn't really matter. Our objective in either case is the same: To work toward a world that reflects God's goals, a world that looks like God wants it to look, a world where there will not only be joy to the world, but joy in the world as heaven and nature (and people!) sing God's glory!

*Joy to the World , the Lord is come!*
*Let earth receive her King;*
*Let every heart prepare Him room,*
*And Heaven and nature sing,*
*And Heaven and nature sing,*
*And Heaven, and Heaven, and nature sing.*

# Reference List and Resources for Further Study

Bahnson, Fred, and Norman Wirzba. 2012. *Making Peace with the Land: God's Call to Reconcile with Creation.* InterVarsity Press.

Ball, Jim. 2010. *Global Warming and the Risen Lord.* Russell Media.

Berry, R. J. 2000. *The Care of Creation: Focusing Concern and Action.* Leicester, England: InterVarsity Press.

Berry, Wendell. 2009. *Bringing It to the Table: On Farming and Food.* Counterpoint Press.

———. 2002. *The Art of the Commonplace: The Agrarian Essays of Wendell Berry.* Emeryville, California: Shoemaker & Hoard (Avalon Publishing).

———. 1990. "Nature as Measure," in *What Are People For?* Berkeley: North Point Press.

Bookless, Dave. 2008. *Planetwise.* Inter-Varsity Press.

Bouma-Prediger, Steven. 2001. *For the Beauty of the Earth: A Christian Vision for Creation Care.* Edited by William A. Dyrness & Robert K. Johnson. Grand Rapids, Michigan: Baker Book House.

Brooks, David. 2010. "The Broken Society." *The New York Times*, March 19, 2010, sec. Opinion. http://www.nytimes.com/2010/03/19/opinion/19brooks.html?src=me&ref=general.

Collins, Francis. 2006. *The Language of God: A Scientist Presents Evidence for Belief.* New York: Free Press.

DeWitt, Calvin B. 2012. *Song of a Scientist: The Harmony of a God-Soaked Creation*. Faith Alive Christian Resources.

DeWitt, Calvin. 1994. *Earth-Wise: A Biblical Response to Environmental Issues*. Grand Rapids, Michigan: CRC Publications.

Diamond, Jared. 2005. *Collapse: How Societies Choose to Succeed or Fail*. New York: Penguin.

Dickerson, Matthew. 2011. *The Mind and the Machine : What It Means to Be Human and Why It Matters*. Grand Rapids MI: Brazos Press.

———. 2010. "Wendell Berry, C.S. Lewis, J.R.R. Tolkien and the Dangers of a Technological Mindset — Flourish." *Flourish Magazine* (online), December 13, 2010. http://flourishonline.org/2010/12/wendell-berry-cs-lewis-jrr-tolkien-and-the-dangers-of-a-technological-mindset/.

Dryden, G.W. 2009. *Farming God's Way Trainers Reference Guide*. Farming God's Way. http://www.farming-gods-way.org/Resources/Translations/FGW%20Trainers%20Reference%20guide.pdf.

Faulkner, Edward Hubert, and S. Graham Brade-Birks. 1949. *Ploughman's Folly*. Joseph.

Hawken, Paul. 2009. "And the Earth Is Hiring". Commencement Address, University of Portland, May 3, 2009. http://cforjustice.org/2009/05/06/and-the-earth-is-hiring/.

———. 2007. *Blessed Unrest*. New York: Viking.

Hawken, Paul; A. Lovins, and L. Hunter Lovins.1999. *Natural Capitalism: Creating the Next Industrial Revolution*. Boston: Little, Brown & Company

Hayhoe, Katharine, and Andrew Farley. 2009. *A Climate for Change: Global Warming Facts for Faith-Based Decisions*. Hachette Digital, Inc.

Hess, Peter M. J. 2006. "Two Books." Encyclopedia of Science and Religion. http://www.enotes.com/science-religion-encyclopedia/two-books.

Jackson, Wes. 1996. *Becoming Native to This Place*. Counterpoint Press.

————. 1980. *New Roots for Agriculture*. New Edition. University of Nebraska Press.

Lausanne Movement. "The Cape Town Commitment: A Confession of Faith and A Call to Action". The Lausanne Movement, 2011.

Little, Brenda. 2007. *The "Green" Gardener : Working with Nature, Not Against It*. Sandy Utah: Silverleaf Press.

Louv, Richard. 2008. *Last Child in the Woods: Saving Our Children from Nature Deficit Disorder*. Chapel Hill: Alqonquin Books of Chapel Hill.

————. 2011. *The Nature Principle: Human Restoration and the End of Nature-Deficit Disorder*. Algonquin Books.

Lowe, Ben. 2009. *Green Revolution: Coming Together to Care for Creation*. InterVarsity Press.

McKibben, Bill. 2011. *Eaarth: Making a Life on a Tough New Planet*. Random House Digital, Inc.

————. 2006.*The End of Nature*. Random House Digital, Inc.

Montgomery, David. 2007. *Dirt: the Erosion of Civilizations*. Berkeley :: University of California Press.

Petrini, Carlo, and Slow Food (Organization). 2007. *Slow Food Nation: Why Our Food Should Be Good, Clean, and Fair*. New York: Rizzoli.

Robert, Dana L. 2011. "Historical Trends in Missions and Earth Care." *International Bulletin of Missionary Research* 35, no. 3 (July 2011): 123–128.

Ryder, Richard Dudley. 1983. *Victims of Science: The Use of Animals in Research*. National Anti-Vivisection Society.

Savedge, Jenn. 2008. *The Green Parent: a Kid-friendly Guide to Earth-friendly Living*. Seattle Wash.: Kedzie Press.

Sorley, Craig. 2011. *Farming That Brings Glory to God and Hope to the Hungry*. Doorlight Publications.

Snyder, Howard and Joel Scandrett. 2011. *Salvation Means Creation Healed: The Ecology of Sin and Grace: Overcoming the Divorce Between Earth and Heaven.* Wipf & Stock.

Speth, James Gustave. 2008. *The Bridge at the Edge of the World: Capitalism, the Environment, and Crossing from Crisis to Sustainability.* New Haven: Yale University Press.

Stott, John. 2010. *The Radical Disciple: Some Neglected Aspects of Our Calling.* InterVarsity Press.

"The Geology of the Planet: Welcome to the Anthropocene." *The Economist.* May 26, 2011. http://www.economist.com/node/18744401.

White, Robert S. 2009. *Creation in Crisis: Christian Perspectives on Sustainability.* SPCK.

Wilson, E. O. 2006. *The Creation: An Appeal to Save Life on Earth.* New York: W.W. Norton.

Wohl, Jessica. 2012. "Wal-Mart Wants PCs to Sleep Earlier." Reuters. New York, September 12, 2012. http://www.reuters.com/article/2012/09/12/us-retail-summit-walmart-idUSBRE88B1GB20120912.

Wright, Christopher J. H. 2010. *The Mission of God's People: A Biblical Theology of the Church's Mission.* Zondervan.

## The Six Principles

**Principle #1:** In a world that reflects God's goals, God's people will be energetically and effectively making God's name known personally and corporately, with activities directed back to God (prayer and worship) and directed to those who do not yet believe (evangelism).

**Principle #2:** In a world that reflects God's goals, all human activity will proclaim God's name by working in harmony with, and contributing to the flourishing of non-human creation.

**Principle #3:** In a world that reflects God's goals, every dimension of society — political, economic, judicial, social, etc. — will operate according to kingdom values like love, mercy and justice.

**Principle #4:** In a world that reflects God's goals, every human being will be able to live a life that is physically healthy and intellectually, emotionally and spiritually rich and fulfilling.

**Principle #5:** In a world that reflects God's goals, every human being will understand that he or she is created in God's image, and will be treated as such a person deserves to be treated.

**Principle #6:** In a world that reflects God's goals, every person will understand that he or she shares responsibility for ruling God's world, and will have the opportunity to exercise that responsibility in his or her own life.

CPSIA information can be obtained at www.ICGtesting.com
Printed in the USA
LVOW122007240313

325757LV00005B/11/P